ANCIENT MYSTERIES, MODERN VISIONS
THE MAGNETIC LIFE OF AGRICULTURE

Also by Philip Callahan
Tuning In To Nature
The Soul of the Ghost Moth
Insect Behavior
Insects and How They Function
The Evolution of Insects
Bird Behavior
Birds and How They Function
The Magnificent Birds of Prey
The Tilma Under Infrared Radiation

Ancient Mysteries, Modern Visions

The Magnetic Life of Agriculture

Philip S. Callahan, Ph.D.

Photographs by the author

Line drawings by Linda Writer

Scanning electron microscope
photographs by Thelma Carlysle
courtesy of the USDA

Acres U.S.A.

Published by Acres U.S.A.
P.O. Box 9547, Kansas City, Mo. 64133
Copyright © 1984 by Philip S. Callahan
All rights reserved. No part of this book may be used or reproduced without written permission except in the case of brief quotations embodied in articles, reviews and books.
Printed and bound in the United States of America.

Library of Congress Cataloging in Publication Data

Callahan, Philip S. 1923—
   Ancient mysteries, modern visions.
   The magnetic life of agriculture.

   Bibliography.
   Includes index.

ISBN 0-911311-08-4
Library of Congress Catalog card number 84-070065

To Charles Walters, Jr.
    My friend and friend of the farmer
and
Hugh Riordan, M.D.
My friend and friend of the sick

About The Author

Philip S. Callahan was born August 29, 1923 in Fort Benning, Georgia. He entered the U.S. Army Air Force in 1943, where he was trained in navigational communications, and assigned to service in Ireland.

After the war, he worked in Japan rebuilding Japan's air navigation system. Later, he was in charge of maintaining radio navigation centers for Japan, Korea, the Philippines and the entire South Pacific. In all, he rebuilt 16 low frequency radio stations. Concerned about the closing of China after the war, he left Japan to hike around the world. While hiking and hitchhiking across Asia and the Mideast, he worked as a free-lance writer and photographer.

Upon returning to the United States, he married Winnie McGee and started college, later earning his B.A. and M.A. from the University of Arkansas and Ph.D. from Kansas State University. He has served in research positions throughout the South and has been awarded with numerous citations for excellence in research. He is the author of some 100 scientific papers and ten books. He lives and works in Gainesville, Florida and remains a world traveler.

Moreover, he has an international reputation as an entomologist and ornithologist, and has been responsible for breakthrough discoveries in both areas. Most important, he is a generalist, and this—his publishers can be pardoned for saying—has accounted for insight and discoveries that arrive only once every generation or two.

TABLE OF CONTENTS

Foreword .... *vi*

For The Record .... *ix*

Prologue — Stones From The Desert .... *xiii*

Chapter 1 — Agriculture and the Round Tower Astronomers 1

Chapter 2 — Agriculture's Electromagnetic Round Towers 17

Chapter 3 — Magnetic Antennae and Ancient Agriculture 26

Chapter 4 — Soil Antennae and Living Auras 38

Chapter 5 — Stone and Clay—The Real Secret of the Pharaohs 53

Chapter 6 — A New Look at Stone 68

Chapter 7 — The Detection of Magnetic Monopoles and Tachyons—A Picture of God 74

Chapter 8 — Sand From East to West 85

Chapter 9 — Monopoles—To Love the Land 97

Epilogue — The Mummy on the Water 111

Appendix 1 118

Appendix 2 121

Acknowledgement 126

Annotated Bibliography 127

Index 132

Foreword

Science, perhaps to an extent unparalleled in any other field of human endeaver, has a very peculiar set of standards, norms, expectations, dogma and even "rules." For instance, freshman science students are repeatedly hammered with the philosophy of the scientific method. The scientific method, a deductive form of reasoning, was designed to provide science with a foundation and framework into which all of the assorted bits of information could fit to form an integrated area of knowledge. It reaches not only into the cataloging and collecting of bits of information, but into actual discovery. It is also drawn upon by scientists to provide a logical way of finding the answer to problems.

The first step in applying the scientific method to the solution of a problem involves carrying out a series of experiments designed to gather all facts about the particular problem being investigated. Then a simple generalization is formulated to correlate these facts. If successful, this generalization becomes scientific law. A jump such as this is seldom made, however, without an intermediate stage—the hypothesis. This is, for lack of a better term, an educated guess. It is one idea that may serve to join the various facts observed. An hypothesis will be subjected to further experimentation in the attempt to find a flaw. If generally unrefuted, the educated guess will earn the status of "theory," where it will likely remain for fear someone will find an exception. It is far more acceptable to disprove a theory than a law.

There is a second scientific method that, although unwritten, has far greater impact on scientists and their findings. This is the reality of project funding, peer review and the publishing of scientific papers. These subjects were discussed in the scandalous book, *The Double Helix,* by James D. Watson, one of the discoverers of the structure of the DNA molecule. He rocked the scientific world by discussing the behind-the-scenes power plays and the jealousy and fighting for funding. But Watson offered these as an aside, showing that scientific discovery is a very human process, not a cold, mechanical ordeal filled with test tubes and microscopes. Discovery relies on a vision.

Like Watson, Phil Callahan has not let the various bureaucracies and administrative tangles taint his love for science and life. Phil Callahan left Louisiana State University at Baton Rouge because he wanted to study biological methods for insect control, whereas the system told him to study pesticides. Many of the discoveries explained in this book are still being bandied about by the scientific community. In fact, Callahan actually has a letter that states, "You went too far," implying that he discovered too much. His discoveries, however, are now a matter of public record, and the right of discovery cannot be denied Phil Callahan. But the implications of Callahan's discoveries are too earth-shaking for a professional journal to risk its reputation in covering.

Phil Callahan does not rebel merely against the formalities of the unwritten scientific method, but against the formal scientific method as well. He openly states that both are ridiculous concepts. A good scientist does not make a discovery through the process of deduction. On the contrary, he discovers through induction. "All great scientific discoveries," says Callahan "originate with observing something in nature. Then you try to explain it." But to utilize inductive reasoning and gather facts from all sources, a scientist must be a generalist. He must be crossed-trained in all fields. Modern practice dictates specialization and fields of expertise. Unfortunately for these scientists, nature knows no boundaries. Her wonders traverse all areas of science.

In earlier times, scientists were literally called "natural philosophers." They were naturalists and generalists. No one seems to ask why the great discoverers were often experimenting in fields outside their area of expertise. They were often great artists, observers of nature, as well as engineers and scientists. For all of their specialization, famous universities, in general, do not produce Nobel laureates. They hire them after the fact.

Phil Callahan is a natural philosopher of the same school as da Vinci, Galileo, Newton and Tesla. His formal training reads of degrees in literature, ornithology and finally a Ph.D. in entomology, the study of insects. He accumulated his informal training as he walked from Japan to Ireland, stopping to observe people and nature along the way. He meditated in the great temples of bygone eras. He lived with the Bedouins in the desert. He ate insects to survive. And he slept in the great stone cathedrals that are now roped off from tourists. It is from this rich background and broad base of experience that he draws far-reaching conclusions that only a cross-trained natural philosopher could account for.

It is exactly because we live in a society of specialization that ancient peoples are misinterpreted. Hieroglyphics in Egyptian temples are misidentified because archeologists do not study entomology. Tales of monks floating in the air are dismissed as folly because anthropologists and historians do not understand natural forms of magnetism. History tells us about wars and death on the battlefield, not about agriculture and the lives of people. Probably the greatest reason for error in our analyses of ancient civilizations is that we relate their lifestyles to our own preconceptions. Phil Callahan is a naturalist, just as ancient peoples were naturalists. He is a 4,000-year-old man who has somehow found himself perfectly comfortable with computerized instrumentation. It is this rare mix that the world does not see enough of, and perhaps it is because men like Phil Callahan are not thought to exist that they frequently go unappreciated.

But Phil Callahan is not unappreciated. Instead of embittering himself to the world and fighting for peer acceptance, he is writing of his discoveries in language that ordinary people can understand, not clouding the wonders of nature with technical hocus-pocus. He is helping us all understand the very nature of life and how it relates to man, his religions and his agriculture. For this we owe Dr. Phil Callahan a very great debt.

Fred C. Walters

For The Record

When I first started to put my *Acres U.S.A.* essays in book form, I thought that perhaps the work should be entitled, *The Transgressions of an Entomologist*. Although that might be a good title for an autobiography, it is not what my essays are about, except to the extent that the stone and soil of which I write have been an integral part of my entomological life for the last forty years or so.

Why have I transgressed? The simple answer to that question is that I have no formal training in either geology or soil, much less archaeology. A good portion of these essays also covers archaeological subjects. In other words, I have no formal credentials to *prove* to my reader that he or she should take anything that I say seriously. In this age of specialization, that is surely a transgression!

As a youth I was, and still am to a certain extent, fascinated by birds. I have a master's degree (a suitable credential) in ornithology. In mid-life I became fascinated with insects. I have a Ph.D. in entomology (also a suitable credential). In my later years I became fascinated by stone and the end product of eroded stone—soil. Although a glimmering of my love of stone goes back to my earlier mountain climbing and falconry days, my intellectual history may be as much a projection of common sense and necessity as of my esoterically-derived motivation. After all, birds move fairly fast and require a good deal of youthful agility to study at hand—especially my main interest, the birds of prey. Insects, on the other hand, are amenable, as befit mid-life, to more leisurely study techniques. Stones are so totally sedentary that any aging person can collect stones from the countryside. Do not be fooled into thinking that, because stones lie silent, they have no life. Stones, in fact, have a secret life quite as fascinating as the *secret life of plants.*

Just as the modern treatises have missed the true *secret life of plants,* so also have they missed the secret life of stone and soil. That secret life involves two little understood magnetic forces called *paramagnetism* and *diamagnetism.*

Stable growing soil is *highly paramagnetic* and all plant life is *diamagnetic.* Those two modern scientific terms may be thought of as what the ancient Chinese called the Yin and Yang of life, the + and − of nature—two equal but opposing forces. The ancients understood these weak but paradoxically-powered forces better than we moderns who—in this age of petroleum farming—have all but forgotten them.

The ancients often used the term *magic* to describe such forces, and invariably worked this knowledge into their religion. The word *magic,* according to Daniel Lawrence O'Keefe in his book, *Stolen Lightning—The Social Theory of Magic,* is found in most languages. It is derived from the word *mana,* which according to that author means "some kind of social power." Mr. O'Keefe, whose credentials are impeccable since he was educated at Oxford, states: "And in many languages the word for magic refers to the same thing—a group of well-known, clearly identified, and unmistakable institutions. The most important are medical, ceremonial, occult, sectarian and black magic."

Mr. O'Keefe's definition of magic is in my opinion incredibly naive, for as my farming friends may note, it does not include the most basic institution of all life—*agriculture.* Incredibly the volume is 580 pages long and the word agriculture does not appear a single time! The sociological reasons for the development of magic is related to every last institution in the realm of human endeavor except the one single institution from which magic sprang and the one single institution closest to life itself—from which all of the other institutions from healing medicine to the perverted black magic sprang.

The very fact that an Oxford-trained scholar could write a voluminous, documented treatise on magic and omit in totality the very basis of all modern civilizations—agriculture—is a powerful indictment of our modern urban way of thinking. It is a way of thinking that is slowly but surely destroying the basic ingredient of agriculture all over the world—soil.

It is also a monument to the incredible ego of high energy, petroleum-oriented man who studies the ancients as mere curiosities, but believes their magic is pure superstition and anathema to modern science. Indeed the word *magic*, from the word *mana*, has been so changed over the eons that today it is associated with subterfuge and not with the incredible Godly force that in other days was utilized to feed, succor and cure the human body and spirit. As a God-created force, it is logical that magic and religion should have gone hand in hand and that the great religious structures of the world should be magical. Baby Jesus was visited by three Magi.

Modern science is a unique, elegant, intellectual system. When used wisely, it can teach us much—not only about ourselves, but also about the wisdom of the ages. Any inquiry into what our agricultural ancestors meant by the term *magic* should be grounded in modern experimental methods.

I, of course, did not discover paramagnetism or diamagnetism. Those forces were first described in the mid-nineteenth century by a group of brilliant German and English natural philosophers—that is, early scientists.

Each of my essays was written as an article for *Acres U.S.A.*, and as such each stands alone. However, there is a common theme that binds them together and that is, of course, my investigations of the magic forces in stone and soil.

The book by Charles Walters, Jr. and C.J. Fenzau, *An Acres U.S.A. Primer*, is a handbook of modern agricultural magic, as is the book just off the press, *The Rest of the Story* by Professor Harold Willis. These are two modern day treatises on how to mix the witches' brew of good soil that will insure the magic of a bountiful harvest. Follow carefully the instructions and incantations in such books, my dear reader, and you will automatically insure that your own farm and garden soil will contain the subtle, but very necessary paramagnetic force described and researched in my *Acres U.S.A.* essays.

Finally I must point out, lest my reader believe that "thou doest protest too much" about my lack of proper credentials as a rock and soil specialist, that I present here the most powerful credentials of all—the credentials of repeatable experimentation. Any objective reader who grinds up a little clay from an old broken flowerpot and then observes that with a 2,000-gauss magnet he or she can attract some of the grains, unrelated to size, but not other grains will come to understand that there is something fundamentally wrong with our modern concept of magnetism.

By understanding these weak, yet powerful stone, soil and plant forces, my farming friends also will come to have a better understanding not only of the ancients, but also of the magic forces in the soil on their farms. I believe I first had a glimmering of those miraculous forces as a young man camping and hiking among a huge pile of stones from the desert called Hueco Tanks. Hueco Tanks was, and still is, a magic place, and my recollections about that place now make a fitting prologue for this volume. The prologue is autobiographical and an introduction to the subject of this book.

Prologue
Stones from the Desert

"The theologian has to restate, laboriously and at length, what is contained in the mystic's flash of intuition."
—*Translator's note, Hymn of the Universe by Teilhard de Chardin*

The sun was a foot above the horizon. It perched like a daisy bloom upon the dried tip of the tall agave plant that grew by the side of the road. I knew that a hike to Hueco Tanks was at best a two-quart walk, for that pile of igneous rocks lay more than six miles across the desert north of the Carlsbad road. A rancher had given me a ride to a spot on the highway where the dirt track of the tanks entered the main road. The dirt track was too insignificant to enter the asphalt highway, so it merely crept up and timidly touched the heat-shimmering black asphalt.

Even though it was late November and a good time for desert hiking, it would probably require the entire two quarts of my canteen to cover the six miles to the tanks. It was at least a two-quart desert walk, even in the winter. The dry desert air accentuates and speeds the evaporation of body moisture. The process effectively cools one, but also quickly depletes body moisture.

I was not worried for I would cover the six miles before the sun got high, and besides, there was plenty of water at the tanks. That is why the place was called Hueco Tanks. *Hueco* is the Spanish word for a hollow rock depression. The wind and water in past eons dissolved and eroded the syenite-porphyry rock where the potholes formed. Each spring the rainwater collects in the pots,

The Canyon de la Virgin at Hueco Tanks. The old Butterfield Stage route lay along the dirt road through the canyon where grass and trees grow in the desert.

Around the canyon walls are huge rock overhangs where Mescalero Apaches once camped.

and the holes in the shaded spots retain the moisture year-round.

Between 1858 and 1861 the Butterfield Stage stopped to water horses and passengers at the tanks. Later they moved the route south to along the current El Paso-Carlsbad road; probably because the Comanches constantly stole horses and mules from the stagecoach station. Contrary to Western lore, the Comanches never attacked the stagecoach itself. During my youth, I believed those exaggerated Western tales, so the trail to the horizon conjured up a vision of the old Butterfield Stage careening along the trail followed by whooping Comanches.

The time was in the early 1940s and there were no Comanches, nor even Mescalero Apaches at the tanks. I was there because I had moved with my mother and sister to my grandmother's big house on Mesa Avenue in El Paso while my father was off at war.

Even though there were no more Indians, something as exciting as a stagecoach, at least for me, occurred on that two-quart hike to Hueco Tanks. A falcon flew by.

The falcon landed on the top of a sotol. It balanced itself for a few moments on the swaying head of the cactus, and then leaped forward and beat its way across the sandy road. Rising high on desert-warmed air currents, it disappeared over the distant mountains. With my grandmother's two-power opera glasses, I barely caught sight of the falcon's rufous and black-tipped tail. Its pointed wings proclaimed it a falcon even though my rather useless opera glasses took in very few of the details of the feathered vision.

I spent the next hour searching agave and sotol cactus for white drippings that would tell me that the bird had a favored lookout perch and had nested somewhere close by. I knew it to be of the desert kestrel variety and very unlikely to have nested anywhere in the vicinity. It was probably a young spring bird that had flown northward from the Rio Grande, leaving the nest. Birds of prey and herons are wont to do just that.

Modern day scientists (who seem to believe that they are superior to the ancients) are convinced they are the discoverers of all such phenomenon, yet I knew in my heart that it could not be so. A few years later I discovered my heart was correct when I obtained Casey Wood's English translation of the emperor Frederick's volume on *Falconry*. Frederick II banned birds in the thirteenth century and traced their migration to and from North Africa. He described in detail the dispersal behavior of herons and birds of prey. My high school books on medieval history outright lied about the medievals. Mankind's mind has evolved insignificantly since those days. I wondered in my heart why modern textbooks felt obliged to put down the discoveries of the medievals. It was almost as if we moderns are so insecure that we negate all the great intellecutal accomplishments of our forefathers. I believe that we applied the same philosophy to the history of the Indians.

Even though I did not find any old nests of the desert kestrel, I could not help but feel excited by the very presence of the beautiful falcon. It was the same magic, trembling kind of excitement that I felt when I once pulled a little screech owl out of a tree hole in Hidden Valley near Detroit a few years before.

I had lost an entire hour from the cool early morning, so it was high noon before I reached the spot on the dirt road where it skirted close to the base of the giant rocks of the tanks. My two quarts of water were gone, but that was of no consequence, for close to the place called *La Cueva del Leon* (the Cave of the Lion) I came to the canyon of Leguma Priety where still water lay in a reed-lined pond.

I knew that life depends on proper water balance and that it takes three times as much water to maintain that balance in the desert as in the shaded, moist woods of the Hidden Valley. Even though the desert was flat, it had taken me more than four hours to hike the six miles. If I had not drunk from my canteen a half a quart at a time in order to saturate my body cells, or if the sun had risen too high, my blood would thicken and begin to coagulate. That would be the end of my two quart hike in the desert. As it was, my blood was as cool and my brain as clear as the moment I left the car on the paved highway. I even had time to pause and listen to the cactus wren calling from the chaparral that hid the mouth of the Cave of the Lion.

When I reached the pond I filled my canteen, and then rolled in the shallow water like a tongue-hanging Indian pony.

By the time I reached the dark, bush-hidden entrance of the cave, my clothes were as dry as the desert air. I wondered if there had really been a mountain lion living in that cave. If there ever was, I had little hope for such an exciting occurrence on that day, for the lions of Hueco had long since been killed by local ranchers.

The syenite-porphyry rock of volcanic origin thrust up into limestone beds less than 65 million years ago. The limestone eroded away leaving the amphitheater-like Hueco Tanks sticking up, an oasis in thousands of square miles of flat desert.

Mountain lions have a developed taste for small, brown calves.

    I searched the high walls of the caves for pictographs left by the artist Mescalero Apaches. I found none, and even doubted if there were any as I had been told by the rancher who drove me along the asphalt highway. Many years later, the tanks would become a state park with a paved road leading to it, and the pictographs would be marked on the tourist map of the rock formations.

xvii

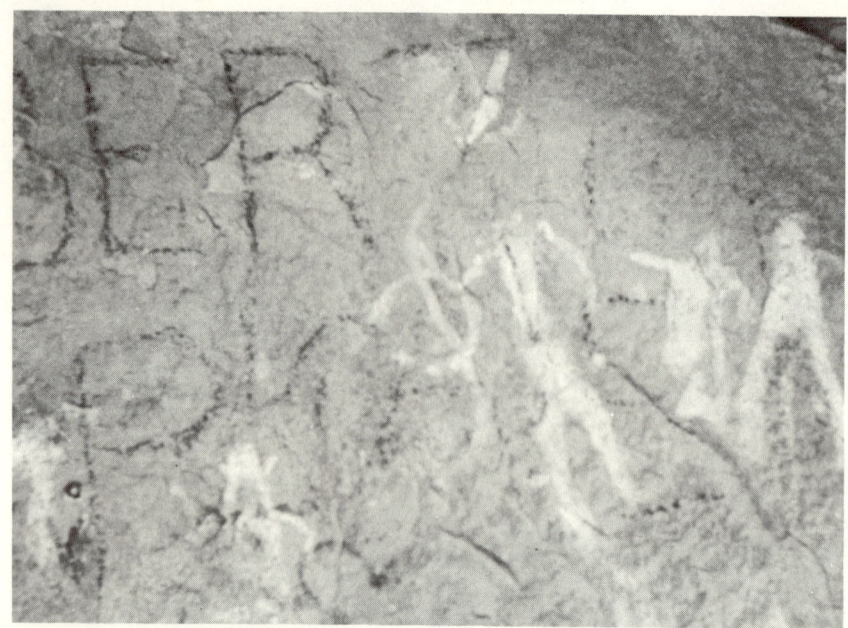

There are three types of pictographs at Hueco Tanks. Black (charcoal) and red (hematite) colors are attributed to the older Puebloan culture. Green (copper oxide) is attributed to the Navaho culture. The white (kaolin or lime) shown in these photographs are the latest and are attributed to Mescalero Apaches. Since these later pictographs depict men in action with horses, they were done after the arrival of the Spanish who introduced the horse to America. Foolish modern man has ruined these pictographs with his egotistical scribbling.

I once had an argument about rock drawings with a friend on a camping trip. My school friend had scraped his initials into a cliff at Red Rocks near Denver. I was willing to fight to stop that desecration, but my friend insisted the Indians drew on rocks, so why shouldn't he. At the time, that logic sounded reasonable and I did not know enough about such things to understand that Indian drawings were visual prayers to a spirit god and also a road map of the souls of those people. Indians did not carve in the rock to proclaim their egotism, *John somebody or other has passed this way*, but to join their own spirit—by way of the symbolic outlines—to their nature gods. Indian picture rocks were not desecrations of God's handiwork, but prayers to heaven. My friend had not understood, and neither had I, so the initial scratching continued. Despite the logic of my friend's argument, and my own ignorance of Indian lore, I seemed to recoil instinctively from pointless graffiti. Such things have nothing to do with art or the soul, but only man's insufferable egotism—or at best a cheap way to obtain some form of immortality by leaving behind indelible identification.

In the Cave of the Lion I found another type of rock graffiti—smoked initials. The floor of the large cave leads, by way of a ledge, to a flat shelf where one can perch. It is a good place to sit and dream the long, long dreams of one's boyhood, or to romance about lives of those desert Indians. I would have lingered for hours, but then I noticed the smoked initials and the magic spell of the darkened room dissipated like the warm whirls of the dust devils that twist themselves into extinction across the desert.

I left the cave and followed the edge of the rocks to the *Canyon de la Virgin.* Hueco Tanks is formed by three large groupings of rocks. One lies east and west and the other stretches out north and south like two huge desert scorpions ready to pinch the mortal sides of the east-west pile with their bifurcated claws.

The Canyon of the Virgin lies between the outstretched claws of the easternmost pile. In olden days the Butterfield Stage passed through the long canyon between the east-west rock and the two scorpion-like formations. The Canyon of the Virgin forms a natural amphitheater to the side of the old trail. It was here that I spent the rest of that magic day among the rocks.

I layed in the sun on the highest peak above the rock amphitheater. I tried to prod my childhood memories of earlier visits to that sacred campground. I could barely remember driving with my mother and father in the old Model A to the Tanks. We often went on picnics with my aunts from the big white house, but the memory of those days was as hazy as that of the broken-beaked hummingbird my father had given me when I was five years old. It had broken its beak by flying against our apartment window. Time is the healer of such injuries, so we fed it from an eye dropper and released it when the beak was long and strong once more. The memory of the incident was so faded that I had to ask my older brother years later whether it was a dream or a real experience. For as long as I could remember, even back to my earliest childhood memories, I had wondered whether or not the mind was really a physical thing—a real part of my brain—or a thing apart. I thought it might be a thing apart for it works in mysterious ways.

In every mind are special memories of quiet places that are truly miraculous earth spots. At such spots, visions and apparitions form. These are, perhaps, human mind apparitions, yet they are just as joyous as God-made ones. The memory of such visions shape one's future life. In my life there was Sand Creek near Denver, where I scrambled down the steep slopes to look for magpie nests; the Guadalupe River near San Antonio where I swam on Texas sunny days; and of course Hidden Valley in Michigan and Hueco Tanks in Texas. There was also Franklin Range which lies like a giant sleeping brontosaurus at the edge of El Paso.

Such are the special places where one may walk or sit at sunrise and sunset among the silent rocks or under strong-branched river oaks. Real earth spots inspirit the soul and enrich our lives, for they are, besides dream places, a catalyst for a special form of reverence where one hears the voices of terrestrial council. The Indians understood earth spots and worshipped the spirits that haunt them. That is why they drew sacred symbols and not graffiti on the cliffs at Hueco Tanks, and why in that place they heard the water spirits singing among the rocks. So also did I on that day of the two quart walk, but during the following weeks the singing spirits would be silenced in America.

The President would call it a "day of infamy," and it would start me on a lifelong journey to many, many lands where spirits sing among the rocks.

NARRATIVE

1
Agriculture and the Round Tower Astronomers

Of late there seems to be a revival of interest in the archaeology of the Celtic people, and it is well there should be, for they were the first Europeans. Over the centuries, the image of that race has been distorted by their Roman conquerors who greatly feared and hated them. Gerhard Hern, in his elegant book *The Celts*, titled his second chapter, "A Roman Nightmare." It documents the wave of Celtic expansion when warrior bands crossed the Alps and temporarily captured Rome in 387 B.C. The term barbarian stuck to those roving people, and as the French historian has written (as quoted by Hern), "The Celts were the most adventurous of all the barbarian people."

Caesar first came up against the Celts on the Iberian Peninsula. Spain had been peopled by the Celts from the sixth century B.C. when tribes from the North (Gaul) had penetrated the Pyrenees. Rome conquered Iberia in the Second Punic War and cleared out the Phoenicians who used Spanish ports to expand trade. As is usual (human nature has not changed much), the Romans represented themselves to the Celts as liberators from the Phoenicians. The Romans, however, declined to go back to Rome, so the army legions soon found themselves engaged in guerilla warfare with the Celts.

The Celts learned warfare centuries before as mercenaries and adventurers—fighting for anybody willing to pay. Caesar's opinion of the Celts was colored by the specter of the fierce Celtic warrior chiefs charging into battle in their wicker-sided chariots with the head of a foe dangling from the neck of one of the chariot's ponies.

Schoolboy history seems to be based on Caesar's opinion of the Celts. What it overlooks, of course, is the fact that between and after the wars, the Celtic tribes invariably settled down to a life of peaceful agriculture.

They were, in fact, Europe's most innovative and technologically-oriented agriculturalists. They developed the technique of metalworking with iron and silver to a fine art. It was the Celts who first put metal blades and the moldboard on wooden plows.

Most of Europe, and two-thirds of Britain, was a hardwood forest biome. There is no doubt that their invention of metal agricultural tools helped the Celtic farmer to clear the massive oak, alder, and ash forests of Gaul and Britain. There is also no question that they invented the first wheeled harvester and also developed an excellent system of utilizing manure fertilizer and crop rotation to insure year-to-year crop stability.

No one really seems to know where the Celtic people originated, but most scholars believe that they came from somewhere east of the Carpathian Mountains, which makes them Indo-European. At any rate, their knowledge of agricultural science was without question based on Eastern techniques. Be that as it may, what we call modern Western agriculture was developed and improved upon by the Celts of Europe and Britain.

The Celtic system of agriculture was totally self-sufficient and practical. Woods were cleared, and small patches of arable land were fertilized with cattle manure and stable litter. During the summer, the cattle browsed on scrub undergrowth and heather. Women spun and made clothes, and the men hammered out weapons and agricultural tools under the soot-blackened thatches of their round huts. The combination of efficient tools and unique concern for the fertility of their soil made for an excellent small farm (tillage-cattle rotation) system that is evident even to this day in certain parts of rural Ireland. This is what Charles Walters, Jr. and C.J. Fenzau, in *An Acres U.S.A. Primer*, would call eco-agriculture—a system we seem to be re-discovering in the face of a persistent loss of our own soil fertility.

The Romans finally drove the Celts into Wales, a corner of Britain, but never did defeat them in Ireland. From the first to the eighth century, when the Viking raids began in force, Ireland was left by and large to itself. It was during the seventh century that most of the elegant round towers of Ireland were built.

Christian missionaries, mostly of Celtic origin and trained in Rome, gradually converted the Irish Celts to Christianity. By the ninth century, Ireland was a land with a monastic network that reached from Horn Head in the north to Cork in the south. The great period of monastic expansion was from the fifth to the seventh century. It was during the end of this era that the round towers were constructed. Today twenty-five or more towers stand upright in perfect form, and the remains, or stubs, of another forty-three dot the countryside. Almost every monastery of any size contained within its protective earthern ring a freestanding round tower.

During WWII, I was stationed at an outpost radio range transmitter site near Belleek, County Fermanagh, Ireland. On nearby Devenish Island in the middle of Lough Erne is one of the best-preserved round towers in all of Ireland. I often rowed a boat over to visit that island for it was a wild and mystic spot—a sort of sanctuary for the long, long thoughts of youth. The tower in Devenish is a finely jointed structure of sandstone. It rises 25 meters above the island and has a base circumference of 15.14 meters. In the fifth, and top, floor are four square-headed windows facing east-northeast, south-southeast, west-southwest, and north-northwest. The lower four floors each have either one or two windows facing in various directions. The doorway is approximately three meters above the ground.

The doorway of round towers is invariably high above ground level. It is commonly believed that this was for defense purposes. Theory has it that the towers were built for protective sanctuary from Viking attacks. Another theory is that they were monastery bell towers. I have always considered both explanations to border on the ludicrous. Large bells were not cast until the Middle Ages (except in China), and the Viking attacks began long after the tower-building seventh century. The monks were hardly that prophetic!

The round tower on Devenish Island, one of the best preserved in Ireland.

It does not take a degree in the history of warfare to understand that as defensive structures, the towers were worthless. It is true that they probably served a secondary purpose as watchtowers for approaching foes in the same manner as a ship's crow's nest.

According to Professor G.L. Barrow, in his book *The Round Towers of Ireland*, the Irish themselves attacked the monasteries more often than the Vikings did in the later centuries. The round towers are like huge smokestacks. A few fire spears or arrows through the lower door or windows would burn the wooden floors and smoke the monks out in no time at all. The towers were not built over springs and could not possibly hold enough food for a long seige. Starvation was the main seige technique in warfare, and we may be assured neither the Vikings nor the Irish attackers were so impatient they would not camp in place for a couple of weeks until the hungry monks crawled out.

If not defensive sanctuary, then exactly why were these towers built? They were not a passing fancy of some impractical mystic crackpots. Not only are they aesthetically pleasing, but over one-third of them have been standing for 1,400 years or more! The stone churches and abbeys came much later in the monastic era.

The author's wife, Winnie, entering the elevated doorway of the round tower at Kinneigh. Common belief holds that the towers were built for protection from invaders. Simple analysis dismisses this theory as ludicrous.

The stump remains of the round tower of Oughterard near Dublin. The round tower originally belonged to a sixth-century monastery founded by St. Brigid. It is made of spalled limestone with a fine granite doorway 2.65 meters above the ground. It is in a graveyard where members of the Guiness family of brewing fame are buried.

Early Irish monasteries were village-size units of wattle huts (interwoven branches and mud) and wooden churches surrounded by a circular earthwork. The Celtic defensive structure was the ring fort. In short, the round tower was the only stone structure in the monasteries during the seventh to ninth centuries. The wooden churches from that era have long since burned or rotted away.

On the inside front cover of Professor Barrow's little *Irish Heritage* pamphlet on round towers is a map of Ireland showing all the still-standing (sixty-eight) towers.

One night, not too long ago, I was lying on my family room couch looking at the map. I kept mentally repeating over and over again, "something is familiar about that map." I would answer to

6   *Ancient Mysteries, Modern Visions*

myself, "Certainly there is. After all it is a map of Ireland." After about five or ten minutes, it suddenly flashed into my mind—insight I believe it is called—exactly why the map appeared so familiar.

The towers formed a star map of the northern night sky. I have used that sky map dozens and dozens of times hiking around in the deserts of the world. It is gouged like a carved woodblock in my brain.

I found a pencil and started connecting the dots on the map.

One of the best preserved monasteries is Clonmacnoise in the center of the great plain of Ireland. It is on the Shannon River and is well known to have been a center for the entire monastic movement. It was only logical to use it to represent the north star Polaris.

To the west of Clonmacnoise I connected the towers of Drumcliff, Dysert O'Dea, Killinaboy, Kilmacduagh, Ardrahan and Roscam and obtained an obvious Big Dipper, Ursa Major. Surrounding Polaris I obtained a series for Draco, the dragon, to the north. To the east was Cassiopeia, the Lady in the Chair, and finally Camelopardalis, the Giraffe, and far to the south Lynx.

There is no mystic astrology inherent in all of this. In fact none of the astrological figures for the months, Virgo, etc., even lie in the night sky at December. I had drawn an almost perfect sky for the December solstice.

The imperfections in the round tower star plot lay mainly in the fact that the monks had to fix their towers to the lay of the land. The Big Dipper should lie west of Draco, not off the tail. The three towers just east of the Galway-Mayo Mountains (shown by an arrow) in the figure on page 9) may have been an attempt to fit the Big Dipper where it belongs. In relation to Polaris, however, it lies more correctly along the rocky edge of the Burren in County Clare where it is located.

There are two Cassiopeias, but the east-west one is positioned horizontally in relation to Polaris. This must have been rectified when a tower was built at Glendalough (see page 9).

Glendalough was one of the most important religious centers in all of Ireland. It was founded by St. Kevin in the sixth century and was the first university in the west.

Round Tower Astronomers

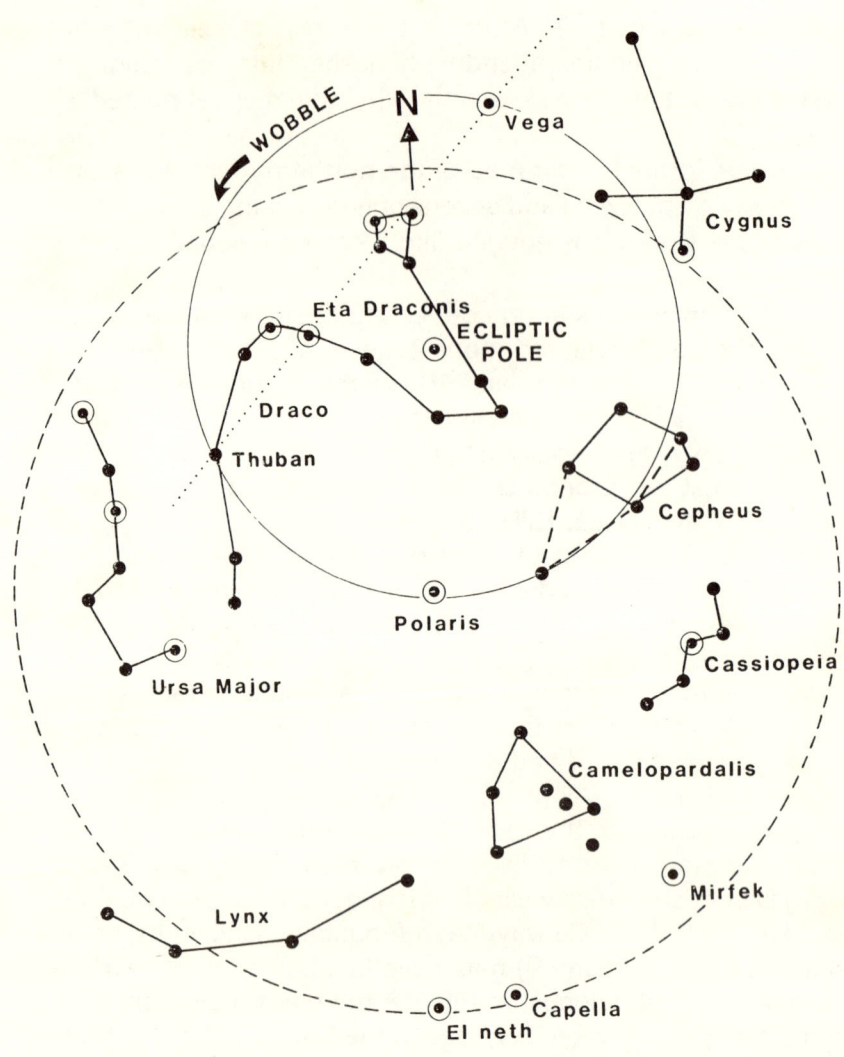

The constellations of the night sky at winter solstice beyond 45 degrees latitude. Polaris is almost directly above the axis of the rotating earth. In the century 2600 B.C., when the Great Pyramid at Giza was being built, the star Thuban would have been the pole star. Thuban is the third star in the tail of Draco, the dragon constellation. A line drawn through Thuban, Eta Draconis, the head of the dragon, and finally Vega, the fourth brightest star in the night sky, gives a straight line across the northern sky that the Egyptians probably used to align the Great Pyramid. About 13,000 years in the future the earth will have wobbled around the ecliptic pole (center) counterclockwise to the point that Vega will be the pole star. The ecliptic pole is of course an imaginary point in the sky. Bright stars are indicated by a double circle.

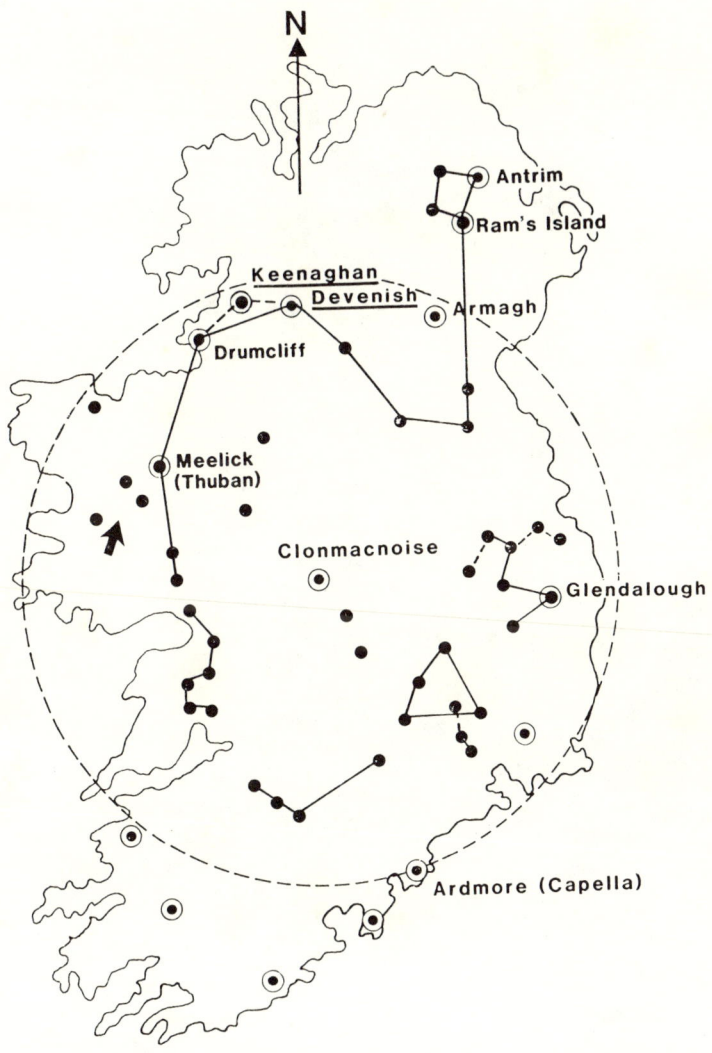

The round towers of Ireland are placed on the ground to match the night sky constellations. Of the four obvious alignments, Draco is the most perfect. Cassiopeia forms two Ws—one west to east and one north to south. Ursa Major (the Big Dipper) is slightly misplaced. Camelopardalis is also close to perfect. The ecclesiastical center of southern Ireland, Clonmacnoise, is the pole star. The ecclesiastical center of northern Ireland, Armagh, is the ecliptic pole. The round tower at Meelick is located at the star of Thuban and Devenish Round Tower, my favorite, is located at the star Eta Draconis. Both were used to align the great pyramid. I am certain that a round tower base could be found at the ruins of the abbey at Keenaghan, completing Draco on the ground. There is a series of round towers along the southern coast of Ireland that make a series of bright stars (first and second magnitude at the southern horizon of the Irish sky).

Round Tower Astronomers

There are as many square towers scattered about Ireland as round towers. They are called tower houses, or castellated houses, and were the fortified residences of the Irish and Anglo rulers of Ireland. They came long after round towers and were of Norman origin, built in the 12th century, between 1180 and 1310. Every Irish tower house tested (15 in number) proved to be built of paramagnetic stone, whereas the cotter or farm houses tested were all diamagnetic.

The most famous round tower in all of Ireland is in the old monastic site founded by St. Kevin who died in A.D. 618. Glendalough is on a point of land between two mountain streams in the beautiful valley of Glendalough. The tower is built of mica schist, one of the most paramagnetic varieties of all stone. Note the offset placement of the windows. Glendalough is considered by Irish scholars to be the first functioning university in the West.

The towers were oriented on the ground in practically the same position as the stars in the northern night sky. In other words, a plot of the towers on the ground formed a rough star map of the northern sky. That phenomenon could be the subject of another entire book, but I must at least mention that most of the northern constellations are outlined on the ground. Of primary interest to this work is the plot of the constellation Draco which is represented on the ground by round towers stretching from County Mayo in western Ireland to Antrim far to the north.

At the tip of Draco's tail is Kilconna in southern Mayo. Next forming the tail are Kielbenna, Meelick (the star Thuban) and Drumcliff (where Yeats is buried). The curve in the tail is between Drumcliff and the beautiful tower at Devenish. The fifth star in line between those two stars is missing on the ground.

If the Celtic monks were purposely imaging the night sky on the ground, then there should also be a tower, or the remains of a tower, somewhere near Keenaghan Lough in County Fermanagh, Northern Ireland. The bend in Draco's neck is marked by Drumlane and Clones in County Monaghan. The head and eyes of the dragon are significantly located on either side of Lough Neagh right in the center of Northern Ireland. On the east side of the large lake two round towers still stand, one located in the center of *Ram's Island* near the eastern shore of the lake and one at *Antrim* in the northeast corner of the lake. There are currently no towers on the west shore to finish the head of Draco, but there are two ancient monastic sites—one at Arboe at the center of the west shore and one on Church Island in Lough Beg, a small lake at the northwest corner of the larger Lough Neagh. An archaeological dig would no doubt find the base of those towers.

The plot also nicely fits the fertile, limestone agricultural lands of Northern Ireland. The drain line of Ireland cuts east of the mountainous, rugged western region of Galway, Mayo and southern Kerry, and south of the highlands of Ulster (the Sperrin Mountains). The formerly forested and present-day agricultural lands, are east of that drain line. The drain line pretty well follows the northwest axis of what geologists call the Caledonia upheaval. It was into the rugged western and northern mountains that the Irish were pushed by the English in the 16th century. The ingenious Irish farmers built soil along the west coast by mixing kelp with what little sand lay among the rocks.

Why is this ground outline of Draco the star dragon so important to my ideas concerning the round towers of Ireland and the ancient pyramids of Egypt? It is important because it demonstrates very clearly that both the Celtic peoples of Ireland and the ancient Egyptians knew not only that the earth was round, but also about the *precession* of the earth and ecliptic center of the sky.

Precession is the technical word used to describe the slow wobble of earth around its axis. The earth spins on its axis rather like a toy top that is slowing down and wobbles around before falling on its side. If one looks from above a wobbling top there is a point in the center of the wobble around which the top moves or precesses, as the astronomers say. That point in the sky above is almost in the center of, and slightly to the side of, the neck of Draco. It is called the ecliptic center.

Polaris, our north pole star, is directly above the axis of our earth at the North Pole so that as our earth spins on its axis it precesses around that point in the sky called the ecliptic center. The slow orbit takes approximately 25,800 years. We see then that 13,000 years from now Polaris will not be our pole star, but rather Vega in the constellation Lyra. Since Vega is the fourth brightest star in the sky, it will be much easier to see than our present-day Polaris. Because we are precessing in space counterclockwise, if we move in time and space clockwise on the orbit for approximately 5,000 years, we come to Thuban, noted by the Meelick round tower, the third star in the tail of Draco. According to all known laws of astronomy, that would have been the pole star when the pyramids were being constructed in about 2600 B.C. Trying to align the Egyptian pyramids with our present-day Polaris is decidedly not the thing to do.

What is astonishing about the round tower star map of Ireland is that there were two great ecclesiastical centers during the early days of Christianity in Ireland, one at *Armagh* in the north and one at *Clonmacnoise* in central Ireland. In relation to the round tower plot of Draco, Armagh is exactly at the point of the ecliptic center. Clonmacnoise is positioned at Polaris which would come quite close to being the only north star 1,300 years ago when the round towers were being built.

The proof that the Egyptians knew about the precession of the earth is carved in stone. On the ceiling of the temple at Denderah in Egypt (300 B.C.) was a circular zodiac that shows Draco as a hippopotamus—they had hippos in the Nile, not Chinese dragons! The hippo's eye falls at the ecliptic center of the zodiac sky. The Denderah zodiac is now in the British Museum for all to see. It is not without credence to speculate that knowledge of astronomy

Round Tower Astronomers 13

and especially of the precession and ecliptic center was carried to Ireland by the Egyptians, and indeed there is a book on that very subject. Along with Scotland, Ireland was originally the land of *Scotus* and Scota was believed to be an Egyptian princess.

I was delighted to observe that my mystic round tower on Devenish Island in Lough Erne ended up being the brightest star in Draco. It is called Eta Draconis.

After I finished my plot, there was still one thing that bothered me—there was one star missing from the uppermost portion of the constellation Draco, west of the star Eta Draconis (Devenish).

Monasteries were usually built at ancient Celtic sites. On my Irish Ordnance Survey map, I plotted the position where the missing Draco star would fall. I knew immediately it was a magic spot. In the history of *The Parish of Cairn* by Reverend P.O. Gallachair appears the following paragraph:

"While many an object of antiquity has been found along the banks of the Erne, then a main thoroughfare of traffic in a thinly populated land (still apart from the old cairn in Dreenan, Boa Island), it is surprising how few remains of the ages of stone and bronze have been recorded from the north side of the Erne here. The southern side, on the contrary, has a large number of ancient settlement sites and burial places of the first people to explore here from the coast. I came across the only ancient site of this kind that is well-known in this district. For a long time it has been well-known to local people that it was some kind of graveyard. Only when I had the pleasure of being allowed to visit it by the Gallagher family who own the property did I realize that here we had an ancient burial place—dating back at least 4,000 years. It contains a number of graves of pagan times, built with the usual impressive, huge stones, and obviously marks a settlement of some of the first people ever to live here. So far, unknown to archaeologists and unmarked on the O.S. maps . . . . This ancient Stone Age site is situated on the property of Thomas Gallagher at Newtown in Keenaghan Townland, about two miles east of Belleek.

It may not be a spot well-known by the Reverend Gallachair, but it was very well-known by me.

Our anti-submarine radio station was not 300 yards from that ancient graveyard site. I could see it from the windows of the hut where we lived in World War II. On more than one day I had walked down to that mystic place and sat among its tombstones and thickets watching the curlews and shore birds that landed along the banks of nearby Keenaghan Lough. Keenaghan is a beautiful little lough that sits astride the border of Northern Ireland.

It was a spot where a 20-year soldier could dream the long, long thoughts of youth; a wild, lonely, magnetic place that collected the winds from the slopes of the Breesy Mountain and rolled them up the Valley of the River Erne.

The farmers I knew 35 years ago from around that lake are just now beginning to die off. Most lived into their late 80s and 90s. The ancient graveyard is called *Keeneghan* by the farmers of that valley.

It was a sacred Celtic spot and it, of course, eventually must have become a monastery. I will be willing to bet that if one were to search carefully the grounds of the old Keenaghan graveyard, buried somewhere at the edge of its crumbling stone wall is the base of a round tower.

One farm family living closest to the spot reared 13 children and who are all still alive and healthy. Their father died at the age of 83 and their mother at 95. They are the heirs of the farming Celts who invented the moldboard plow. Even today the oldest son farms 40 acres with much the same tillage-cattle rotation system as did his Celtic ancestors. That farm has had no land erosion problems and its owners live happy lives—but then why shouldn't they if they are near a mystical round tower? The Reverend Gallachair says of the parish of Cairn (as quoted from an 1835 work), "the tract of country is healthy in an uncommon degree, so that a physician or even an apothecary will find it impossible to support himself by his practice."

If the German geobiologists are correct and there are good zones and bad zones (geopathogenic) on the earth's surface, then most assuredly those Celtic monks built their stone observatories at earth's harmonious points. At any rate, I intend to find out.

I have learned how to plot the resonant waves from thunderstorms by utilizing a potted fig tree as an antenna hooked to an extremely sensitive electrometer. The same technique might also succeed with coherent waves from round towers' resonant places, if they exist.

The technocrat, who is high-energy, inorganic-slanted, will of course scoff at my star map of round towers and say that the correlation is coincidental. For the high-energy technocrat every phenomenon that does not "hit" one on the head with an inorganic "hammer" is a coincidence. Coincidence is the cop-out word of the century used to put low-energy organic researchers in their places.

The best proof that the round towers were laid out as a true map of the night sky is the simple fact that the map of the sky imprinted in my "desert hiking" brain correlated in my mind with Professor Barrow's round tower map in a microsecond flash! Yet if a computer had made the same correlation, it would immediately be accepted as a fact. That unfortunately is what has gone wrong with present-day research—more faith is placed in high-energy inorganic technology than in God-made organic systems. But whether or not I am correct, this book will at least provide an excuse for an old soldier to return to Ireland to dream long thoughts about his Celtic ancestors.

2
AGRICULTURE'S ELECTROMAGNETIC ROUND TOWERS

In the year 1884, a tall, gawky youth landed in New York City having emigrated from his native Yugoslavia. The world has never been the same since, for Nikola Tesla was the true father of our electrical industrial revolution. By the beginning of the twentieth century, Tesla had invented the AC motor, the radio (the resonance circuit), the fluorescent tube, discovered superconduction, and possibly built a laser. The list of his many other inventions is too long to relate here; suffice to say that while Marconi was still struggling to generate static electricity, Tesla had invented a radio-controlled boat and was guiding it around a pond in Madison Square Garden before hundreds of electrical engineers. Marconi is given credit for inventing the radio, but he did not. The popular history of science is more distorted than the history of the American Indians with all its historical myths.

Along with Thomas Jefferson, Ben Franklin, Charles Lindbergh and a few other really great Americans, Tesla is one of my heroes. He was a true American. His notes and patents from before 1906 were destroyed when his laboratory in New York City caught fire, but not his citizenship papers. He kept them in a fireproof safe!

Tesla was, without any doubt whatsoever, responsible for much of what we Americans call our "high standard of living." Despite my admiration for Tesla the genius, I must constantly stop and ask myself, "Have we carried his electrical revolution too far?" By that I mean, must we continue with the mistaken notion that every advancement of civilization be based on the utilization of inorganic high-energy systems? I have come to believe that our constant insistence on generating more and more energy is analogous to the person who increases his intake from one afternoon cocktail to five per day. We have become high-energy drunks and we should pause to think upon whether or not this is in harmony with the human evolutionary process.

I shall not use my space in this book to emphasize the dangers inherent in nuclear energy. After Three Mile Island, these dangers should be obvious. Instead I shall restate my belief that we must utilize our talent re-learning how to "tune in to nature." I will, in fact, go further and state my belief that our agricultural ancestors—the Celts—already knew how to tune in to nature and that we, through historical circumstances, have forgotten their methods. The reasons for our collective forgetfulness are too complex to explain in fewer words than an entire book might contain. Be that as it may, I am certain that the primary reason that we do not revive some of those ancient organic systems is our own technological arrogance. We have made a god out of high-energy inorganic technology in somewhat the same manner that our Egyptian and Celtic ancestors made a god out of low-energy organic systems.

To understand exactly what I mean, you as my reader must understand the fundamental difference between our high-energy way of thinking and the ways of the ancients. A case in point is the body of knowledge (some would not label it so) called astrology. We cannot, by present standards, call astrology a science—for it has been perverted and exaggerated by modern opportunists. However the fact that ancient astrology had merit was, in fact, easily demonstrable.

In April 1946, John Nelson, an electrical engineer with RCA, was asked by his supervisor to investigate the sun—a strange request for an electronics specialist. Nelson obliged by setting up a telescope on the RCA Building in Manhattan. RCA was concerned about sunspots which interfere significantly with telecommunications, and the firm wanted to be able to predict that interference.

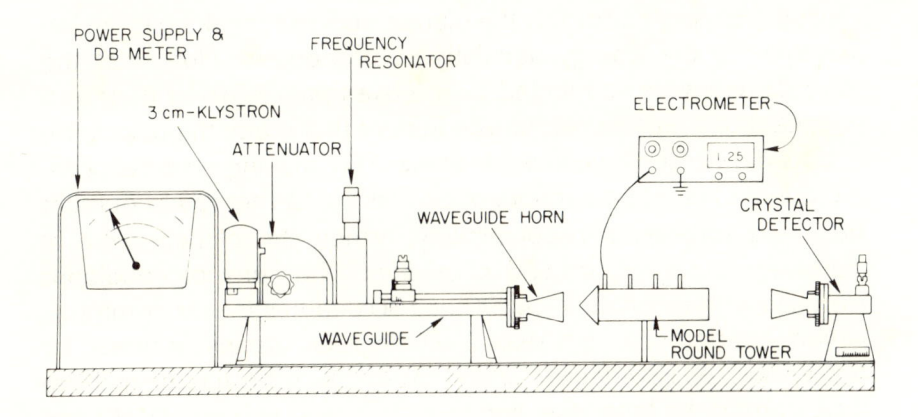

This drawing shows a sandpaper round tower model in the beam of a 3-centimeter radio emitter. The klystron puts out the 3-cm. wave which is guided by the waveguide horn to the top of the model round tower. The electrometer can be attached to the tower at different levels. The digital meter (or recorder) on the electrometer reads out current flowing on the surface of the tower. The same signal in the 3-cm. region is obtained from the sun or is obtained from this man-made system, demonstrating the round tower configuration can collect and detect radio waves from the sun (see appendix 2).

After two or three years of devoted research, Nelson discovered something very peculiar. He found that whenever the sun was positioned exactly between Mercury and Jupiter—0 degrees and 180 degrees on either side—and when Mercury was moving opposite Saturn, and also Venus was 90 degrees to Saturn, the incident of large sunspots increased.

This planet-sun orientation gives two angles of 180 degrees—Mercury opposite Jupiter and Saturn opposite Mercury—and one angle of 90 degrees—Venus at right angles to Saturn. The orientation occurs during a four-day period.

RCA released the results of Nelson's astronomical observations. Up to that time, astronomers were firmly convinced that the weak energies emitted by planets could have little effect on anything in the cosmos. After all, the billions of suns emit mind boggling amounts of both short and long wavelengths of electromagnetic radiation, and in terms of gravitational pull, the earth is nothing to

Electromagnetic Round Towers 19

the sun. If we believe that our own sun is powerful, then consider the Crab Nebula that radiated at a rate some 100,000 times that of our sun (radio through X-rays).

After Nelson's discovery, the planets suddenly took on more importance to the energy-calculating astronomers. Note that the astronomers were so blinded by the fixed paradigms of their own science that it was an electrical engineer that made the discovery.

How did Nelson calculate his angles? By making up a heliocentric-oriented (sun) horoscope, of course. He further found out that he could forecast sunspot activity. When the planets were at 120-degree angles, instead of aligned, the sunspot conditions were likely to be good. Zero, 90 or 180 degrees means communication interference, whereas 120 degrees means minimal interference. A 120-degree angle called the trine has been from time immortal a favorable sign to astrologers, whereas 0, 90 and 180 degrees have been considered a forecast of ill omen. Nelson's work was a triumph of low-energy thinking (planet effects) over high-energy thinking.

Enough of our contempt of the ancients and their science of astrology—let's proceed to the Irish round tower astronomical observatories that I wrote about in Chapter 1.

When I was growing up and building my own crystal radio receiver, I remember wondering how that little rock crystal converted those nebulous radio waves to a sound that I would hear as music or speech.

I later learned that the crystal radio receiver is a low-energy resonant system for collecting radio waves. A coil and wire of proper length act together to collect the weak radio waves, and the cat's whisker and crystal operate as a wavelength detector. We will not go into the complexities of solid-state physics, but a detector is a component of electronics that takes the electromagnetic wave from the antenna coil and converts the wave to DC (flat) electrical current—this is called rectifying the current. Essentially, when the radio signal is strong, the rectified DC coming from the detector goes up, and when the signal is weak, the rectified DC current from the detector goes down. That, in simplified terms, is how a radio receiver works. There are only two really essential components—the antenna coil, cut and wound to match the incoming radio wave, and the detector rectifier—to rectify the current.

Early in 1904 it was discovered that crystals of galena and carborundum both had excellent semiconductor rectifying properties. Carborundum is a man-made crystal of silicon carbide. It is almost as hard as a diamond. By touching a fine wire from the antenna coil (called cat's whisker) to such a crystal, the alternating radio waves on the antenna are rectified.

Silicon is not only a good DC rectifying substance but it also has other useful properties. It is used in solar cells to convert light energy from the sun directly into electricity. Later, during World War II, it also was discovered that at high frequencies—centimeter and meter wavelengths—vacuum tubes did not make good detector rectifiers. The need for high frequency detectors for radio during the war stimulated researchers to go back and look once again at crystal detectors.

A Bell researcher, George Southworth, began searching secondhand electronic shops for obsolete silicon detectors. Most solid-state systems (transistors) are based on his work. Like Tesla, he never received a Nobel prize. Others took credit after building on his original research.

George Southworth instigated another piece of elegant research. In 1945 he discovered centimeter wavelengths being emitted by the sun. He found that 1-10 centimeter radio radiation moved across the sky with the sun.

The first discovery of radio waves from the cosmos was made by Bell's Karl G. Jansky. As reported in one of the classic radio papers of all times, titled *Directional Studies of Atmospherics at High Frequencies,* in 1932 Karl Jansky discovered radio waves at 14.6-meter wavelengths coming from the night sky.

Interestingly enough, a photograph of Dr. Jansky shows him by his recorder sketching wavy lines across a star map of the night sky in the region of Cassiopeia and the North Star. His map matches exactly my own plot of the Irish round towers on the ground as discussed in Chapter 1. Even more interesting, the round tower at Devenish has wavy lines carved into the stonework directly below its conical cap—Celtic electronic sine waves, so to speak. The Celts were addicted to carving wavy lines in stone! In light of the above, we must ask ourselves if the round towers of Ireland might not really be huge resonant systems for collecting and storing meter-long wavelengths from the cosmos. If so, for what reason

Electromagnetic Round Towers 21

and how would such a tower function since it is not constructed of metal like a radio antenna?

Obviously round towers are not conventional antenna. They are in fact built of limestone or sandstone blocks and are therefore closer to silicon semiconductors than to metallic conductors. I believe that the towers have not only the properties of a DC rectifier, but also the ability to detect and store incoming cosmic electromagnetic/magnetic energy, thanks to their dielectric properties.

At high frequencies, the best antennae are not made of metal conductors, but of dielectric (insulative) substances such as plexiglass, wax, etc.

The physics of dielectric systems is extremely complex; suffice it is to say that they can be formed into tubular or rectangular waveguides to collect and guide energy in the same manner as a metallic radio or TV antenna. Since round towers are meters high and meters in diameter, they must be collectors of meter long cosmic radio wavelengths or magnetic energy.

The problem with studying low-output energies from dielectric systems is that it is extremely difficult to measure them. Therefore, most scientists make no attempt to do so, and those who do often find themselves scratching their heads. Fortunately, if one really understands resonance ("tuning in to nature," as I call it), then it does not take a lot of money or equipment to study such systems. The most economical method is, of course, by modeling—a method well-known in the aircraft industry. It is much cheaper to study the aerodynamics of an aircraft model than it is to build an aircraft and then find out it will not fly.

It was fairly simple for me—using the dimensions published in Professor Barrow's book, *The Round Towers of Ireland*—to construct accurate models of round towers. The question, of course, is to construct them of what? After a little thought the answer popped into my head—insight, I believe it is called. Why not use plain old sandpaper since the tower is constructed of sandstone? Better yet, why not use metal polishing paper since it is made of carborundum (silicon carbide) of crystal set fame? I would then have not only a round tower antenna, but also a rectifier all rolled into one.

I bought a few sheets of carborundum paper, investing forty cents per sheet, and then looked up the properties of silicon carbide in the 1954 edition of the *Encyclopedia Britannica*.

The first line of the write up of silicon carbide was very disappointing to me. "Silicon carbide is one of the small class of solid compounds containing only non-metallic elements." If silicon carbide is absolutely pure, then how could it possibly be a semiconductor substance? Semiconduction in dielectric substances depends on trace element impurities—especially of metal or rare earths. Physicists call it "doping." I call it "a little is a lot." Our health depends on doping, vitamin C, iron, etc.—good; cigarette smoke, marijuana, etc.—bad!

A little further along in the encyclopedia I read how silicon carbide is manufactured. Extremely high current (20,000 amps) is passed through carbon electrodes into a mixture of silicon sand, coke, sawdust and salt. Sawdust, coke and salt? How could it not be doped with traces of the many metallic and rare earth elements contained in plant material, and also salts of various kinds?

Rare earth metals and certain salts are known to have paramagnetic properties. Most substances are either diamagnetic, meaning repelled by a strong magnetic field, or paramagnetic, meaning attracted.

Paramagnetism describes substances that have a type of electron configuration in their atomic orbitals that makes them weakly magnetic.

I built my model towers to the exact dimensions of the tower on Devenish Island in County Fermanagh, Ireland. Utilizing a high-frequency oscillator called a Klystron, I generated three-centimeter wavelengths of radio energy. Devenish round tower, which is aligned to the night sky, is 25-meters high, so it should resonate to meter-long wavelengths like those discovered in the night sky by Dr. Jansky in 1932. I reasoned that my centimeter-high sandpaper tower should resonate to the centimeter wavelengths from my Klystron, or the sun (see figure, page 19).

When I put my ten-centimeter (three-centimeter diameter) sandpaper tower in the radio beam, the power meter went up from 6 DB of energy to 9 DB of energy, proving beyond any doubt that round towers are radio waveguides. Just as a glass lens will collect and focus light making it brighter, so did my round tower in the case of three-centimeter radio waves (see appendix 1).

The question remained, however. Could a little sandpaper tower not only collect the energy, but also detect and rectify the wavelengths? I set the tower upright and connected it to a sensitive electrometer and a continuous chart recorder.

Dr. Southworths' 1945 paper titled, *Microwave Radiation from the Sun* shows a flat line at night, but during the daylight (sun) a gradual hump or increase in power proportional to the swing of the sun from the horizon to a noontime overhead position. His data also shows many small waves or oscillations during the day. I received the exact same signals with my round tower model connected to my electrometer. A silicon round tower is indeed an antenna waveguide and silicon rectifier all rolled into one. There is absolutely no reason to believe that if I can collect and detect cosmic radio with a 40 cent piece of doped, paramagnetic sandpaper (silicon), that a 25-meter tower of doped, paramagnetic sandstone will not in the same manner tune into cosmic meter long wavelengths from the night sky (see appendix 1).

In his book on the round towers of Ireland, Professor Barrow makes the point that documentary sources on the towers are few and unreliable. The only thing certain—as in the case of the Great Pyramid—is that they exist. One of the most controversial questions concerning the towers is the high placement of the door in each tower. The high doorways do, of course, have a security function where common robbers are concerned, but as stated before, the assumption that armed troops, Vikings or otherwise, would be defeated by such a defensive mechanism borders on the ludicrous. Could there, perhaps, be another more electromagnetic reason for the door placement? There is indeed. No matter how mathematical electrical engineers pretend antenna design to be, it is in reality experimental and empirical. First come the design and measurements; then when one finds out all the figuring does not assure sharp resonance, out come the wire cutters to shorten the antenna or more wire to lengthen it, or perhaps a little bending, curving or twisting until the decibel signal comes in loud and clear from the transmitter.

It would, of course, be extremely difficult to stretch or shorten a round tower—or would it? Not if the doors were placed high enough! According to Professor Barrow, one of the strangest things about towers is that the space between the door and the ground is filled with dirt. The filling, depending on the tower, varies from none to all, clear up to the wooden floor at the doorway. A reason given for the dirt filling is that it strengthens the base. This reason contradicts good mechanical principals of construction. The force of the mass of the tower walls is downward with gravity. Filling the interior with packed dirt would put an outward force against the wall right at the base where it would be least desirable!

The monk round tower builders could easily tune the tower to the night sky radiation by going inside and filling the base with dirt until they received the right message. The question at this point is, "What is the right message from the cosmic sky?" Modern man has been led to believe, since most microwave radiation is man-made, that it is unnatural. The simple fact is that man and all of nature's plants have evolved over billions of years under low-energy microwave radiation from the cosmos. It is a type of low-energy radiation that constantly bathes not only our own bodies, but all of our agricultural crops.

The monks were past masters at meditation, and meditation—as William Johnston points out in the title of his book—is a form of *Silent Music*. The most successful agriculturalists of all times, George Washington Carver, Luther Burbank and many other great plant scientists have documented the "silent music"—although not in that exact term—that they received from their plants. The Irish monks were the first of the West's great agriculturalists and they may well have known how to "tune in to nature" not only within themselves, but also for the benefit of the fertile Irish soil.

There is nothing in the encyclopedia about magnetic antennae, but we have all read about research on the effect of magnetic fields on seeds and plants. How does it work? Next, I will demonstrate that not only are these structures massive electronic collectors of cosmic microwave energy, but that they are also giant accumulators of magnetic energy. They are, in other words, "tuned" magnetic antennae, and as such, most certainly contributed beneficial cosmic energy to the fertile fields of those ancient, low-energy, stone, electrical engineer monks.

3
MAGNETIC ANTENNAE AND ANCIENT AGRICULTURE

"Let me learn the lessons you have hidden in every leaf and rock."
—*An Indian Prayer*

Before discussing Irish round towers as a magnetic antenna, a short recapitulation is in order. Earlier I discussed the history of the Celtic people and the elegant round towers scattered around the countryside of Ireland. I showed that they most certainly were not refuges from attack by Viking armies, nor was their main purpose for the hanging of large casts bells or bell towers.

The towers were constructed at ancient Christian monasteries where the Celtic monks practiced a form of eco-agriculture dependent upon crop and pastoral animal rotation. Indeed, these ancient Celtic people were the forefathers of good, modern, biological farming.

Following this, I demonstrated, by a modeling technique, that the Irish round towers are in all probability huge, well-designed, stone waveguide detectors of microwave radiation from the cosmic universe. I arrived at that conclusion because—as I pointed out previously—the towers are aligned with the stars of the night sky at winter solstice and we know that cosmic microwave radiation at 14.6-meter wavelengths is emitted from that region of the universe. I detected centimeter wavelengths from the sun with model, centimeter, carborundum towers.

You will also remember that I pointed out that despite what the encyclopedia says, carborundum is not a pure substance—silicon carbide. It is most likely doped with rare-earth metals or salts, since it is formed in the presence of sawdust. Sawdust, of course, is a plant material and contains within its grains traces of all sorts of elements. I decided, therefore, that carborundum paper, which is a good substitute for sandstone, might be a paramagnetic substance.

A paramagnetic substance is a non-metallic compound that has an electron configuration in its atomic orbital that makes it weakly magnetic. Since round towers are mainly constructed of sandstone and mortar, they probably have low-energy paramagnetic properties. If so, the question we must ask ourselves is, what part does the characteristic slope of the tower play in its paramagnetic properties? In other words, are the towers a magnetic antenna? Does their efficiency depend upon their form? Keep in mind while reading this book the slope of the Indian tepee, the witch's hat, and also the dunce hat of the Victorian schoolroom.

My carborundum round-tower models worked as I predicted, but the final test of such a scientific project is to look into the towers themselves and attempt to detect electromagnetic or magnetic energy. I plan to do just that in the near future, but in the meantime, I must collect all of the information possible on the subject. One of the weaknesses of science today is the poor literature review, despite computers, that often proceeds such scientific projects. The earliest writings are most often ignored. This is also a human factor indicative of our contempt for the ancients.

In my literature review, I discovered that of the sixty-four standing towers, twenty-five were built of limestone and thirteen of sandstone. The rest were built of basalt, clay slate, or granite—all minerals that we might expect to have weak paramagnetic properties. Much of the sandstone in Ireland is red sandstone, indicating a low concentration of iron. Limestone is composed mainly of calcite, but impurities of clay are abundant in it.

Earth clays contain traces of iron. For instance, ordinary clay flowerpots are paramagnetic. I often have heard plant growers say that they thought their plants grew better in clay flowerpots than in modern, plastic flowerpots. Despite the fact that red clay flowerpots are paramagnetic, and thus weakly attracted by a powerful magnet (1,000 gauss or above), I could not find any work other than my own (unpublished) on that physical characteristic of clay flowerpots.

If we look in *Dana's Manual of Mineralogy*, we find that sandstone is a sedimentary stone composed of grains of sand bound together by carbonate, iron oxide (hematite or goethite) or argillaceous (clay) materials. Hematite (Fe_2O_3) is a natural magnetic substance widely distributed in rocks all over the world. Hematite or magnite mixed with the mineral corundum (Al_2O_3 is called "emery"—which, of course, means that emery sandpaper and emery fingernail files are also paramagnetic. Likewise, aluminum is a paramagnetic substance. The paramagnetic, iron-doped compounds, such as emery, are usually labeled "ferromagnetic," but we won't worry here about the subtle differences between para- and ferromagnetic compounds, since both exhibit weak magnetic forces. I construct my model round towers from both carborundum and emery sandpaper.

As we mentioned before, a characteristic of a paramagnetic substance is that it is weakly attracted to a strong magnet. In my work, I use a 1,000 to 2,000 G. samarium cobalt-alloy magnet. "G." is the symbol for "gauss," the measure of a magnetic force. It is also called an "oersted." In simple terms, one gauss is the unit of magnetic induction that will induce one volt in a centimeter length of wire moving one centimeter per second through the magnetic flux. We need not go into the complexities of measuring the extremely weak paramagnetic forces of emery or carborundum sandpaper, but no complex equipment is necessary.

A model round tower suspended by a thread one quarter-inch from a strong magnet (1,000 G. or more) will orient with the conical point toward the magnetic field. So will an emery fingernail file suspended by a thread or a small triangular piece of broken clay flowerpot since they are also paramagnetic substances. However, because it involves weak magnetic forces, you will need to hold them close to the magnet.

If you cut a small piece of emery or carborundum sandpaper into the form of a witch's hat—an elongated cone—and suspend it by a thread from the side, you will find it to be highly attracted to a magnet across the base of the cone opposite the point.

Should you move the cone very carefully with no twist in the thread and with the point exactly vertical to and toward the

28 Ancient Mysteries, Modern Visions

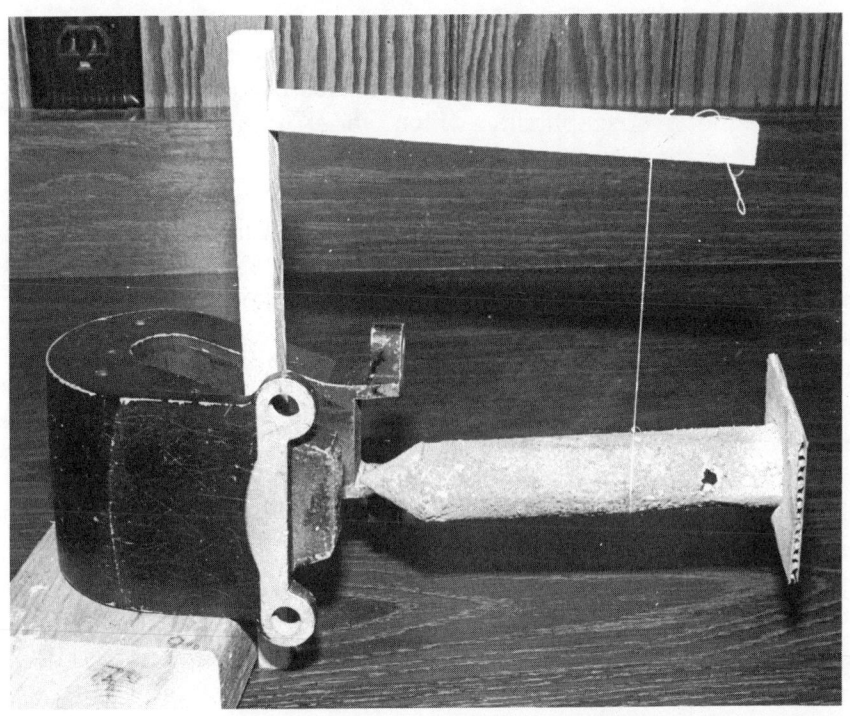

Model of Devenish round tower made out of a rolled-up five-by eight-inch index cards and coated with crushed grains of a red clay flower pot. The granules alone will not stick to the 1,000-gauss magnet, but are attracted to the magnet when formed into the shape of a round tower or, as I call it, a magnetic antenna. Devenish is the only round tower in Ireland that has wavy lines etched into the wall stone just beneath its conical roof.

magnet, then the point will attach to the surface of the magnet. If, however, you allow the "witch's hat" to turn away the least bit from the vertical, the magnet will pull the base around, like a weathervane in the wind, and draw the broad base to the surface of the magnet. In the case of a conical witch's hat or dunce hat, the base will be the portion sitting on a person's head!

A piece of flat emery paper, since it is more highly paramagnetic than a piece of flat carborundum paper, will move toward the magnet if suspended lightly between the thumb and forefinger. A piece of flat carborundum paper so held will not be attracted at all to a 1,000-G. magnet. However, as soon as the carborundum paper is rolled into the shape of a round tower with its conical shape, it also will be drawn to the poles of a powerful 1,000-G. magnet. It becomes a magnetic-formed antenna.

Magnetic Antennae and Ancient Agriculture

At this point, we must define a second property of a paramagnetic substance. Such a substance has by definition a fixed, weak-magnetic susceptibility which varies only slightly with a magnetizing force.

Most of us realize that if we take a steel screwdriver and rub it against a strong magnet, the magnetism will transfer to the screwdriver. That is because substances such as steel and cobalt have high magnetic susceptibility—they are magnetic metals. The harder and longer we rub the screwdriver against the magnet, the stronger the magnetic energy it "absorbs" into its own atoms. This is not true of a paramagnetic substance. Rubbing a clay flowerpot against a strong magnet will not increase the "absorption" of the magnetic energy into the pot. I put the term "absorption" in quotations marks because despite all of our field theories, we still do not understand magnetism very well at all. We simply do not have a good understanding of how the energy transfers or, for that matter, what magnetism really is. We do understand that magnetic energy can be an antigravity force. A strong magnet will support another magnetic substance in space. We know even less about gravity. Physicists for the last 20 years or so have been attempting to measure the gravity waves that Einstein predicted. They have not succeeded!

Another experiment we can try is to sprinkle iron filings on a piece of paper over a magnet and observe the lines of the field forces. The magnetic energy orients the filings in lines between the poles. This brings up the question of how one can study the lines of force around a paramagnetic round tower.

In order to strengthen my contention that round towers are paramagnetic antennae, I had to be able to see field force lines on my models. The question I asked was, are the forces projected in waves around my model paramagnetic towers? Obviously, I could not use iron filings to give a field force pattern. There simply is not enough energy to move iron filings around. I had to use a substance light enough so that the small, model towers could orient the grains. Aluminum, you will remember, is a paramagnetic metal, but even aluminum filings are too heavy to be moved by an emery or carborundum tower.

I had a idea—insight, I believe it is called—why not use Epsom salt? Magnesium sulfate, unlike aluminum, is a diamagnetic

Two carborundum round towers. The tower at left is modeled after Devenish round tower and the one at right after Turlough round tower. Both have been soaked in a solution of Epsom salts for 24 hours and dried in the sun for 48 hours. Note the very fine field lines of concentrated salts around both towers. On the Turlough tower the salts concentrated heavily at the levels where floors and windows are located.

Magnetic Antennae and Ancient Agriculture

substance and Epsom salt is a form of magnesium sulfate. It is a light, white powder and the small grains might be moved by weak forces. Furthermore, like sandpaper, it is easily obtained from the corner store, in fact, in some cases, right from a home medicine cabinet.

I knew I would have to dilute and spread the salts in water over the round tower surface. It probably would not work on the stronger paramagnetic emery paper towers because the emery surface is too rough. Carborundum paper, however, is used to polish metal and is coated with a thin plastic layer so that a jet of water or oil (for cooling the heat of friction) won't soak the paper and ruin the sheet of carborundum.

I modeled the short, squat round tower located at Turlough in County Mayo, Ireland. It is 21 meters high and 17.5 meters in circumference. The short, squat configuration makes it one of the few deviations in size among round towers, however, a centimeter model of that dimension fits nicely (cone down) into a Mason jar filled with cold water in which I put six tablespoons of good old Epsom salts (see appendix 2).

I left the carborundum paramagnetic round tower soaking 24 hours, then took it out to dry for 48 hours.

Sure enough, when I examined the model closely it was circled with thin, white lines about one millimeter apart from the base to the top. The tower cap or cone also was circled with lines in a helical pattern that spiraled to the base. A straight line on a flat surface automatically forms into a helix when shaped into a cone. The lines of force spiral from the tip to the base.

Even more astonishing, at each level on the model where a floor and window would be located on the Turlough round tower was a broad, strong band of white Epsom salt! In other words, the towers are indeed magnetic antennae used for concentrating paramagnetic energy. These lines of force are quite similar to the standing waves of energy that can be measured on a resonant electromagnetic radio antenna. Such standing waves of concentrated energy are called electromagnetic modes by electrical engineers. What I have discovered are low-energy paramagnetic "modes." In other words, the towers are designed so that the strongest "modes" (lines) are concentrated at the floors of the towers where the monks would be sitting and chanting or observing the stars

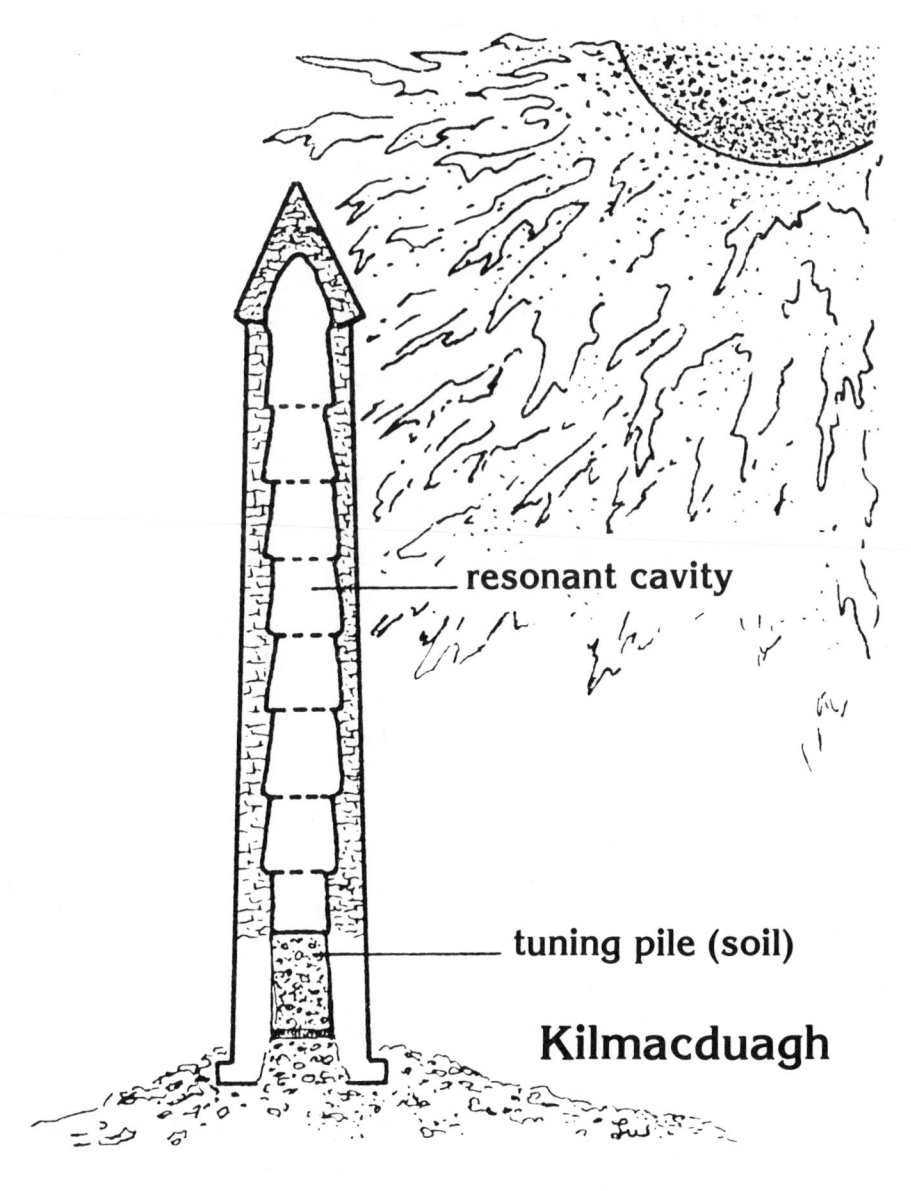

The round tower at Kilmacduagh in County Galway, Ireland is the tallest of the towers, being 34 meters in height. Part of the base below the doorway of such towers was often filled with dirt. This could well be a dirt tuning pile for shortening or lengthening the rock antenna.

Magnetic Antennae and Ancient Agriculture

through the little tower windows.

There is an aesthetic joy in discovering one of the secrets of nature that transcends any other worldly accomplishment. Also, there is additional satisfaction when such discoveries are made with simple, dimestore experiments not requiring thousands of dollars worth of grant money. That, of course, is why I say that true science is not technology, but the contrary—real science is poetry. By that I mean that the competent scientist is the person who can watch the stars and measure their beauty rather than watch the stars to measure their light. Science has the elegance of poetry, for science lies deep within the soul and never deep within the halls of technology.

What do paramagnetic round towers have to do with farmers and farming? That is, by far, the easiest of all questions to answer. Although there is very little research directed at the effect of magnetism on agricultural crops, there is enough to give us some very interesting clues.

On one of my trips across Lough Erne to Devenish Island where the Devenish round tower is located, I asked the fisherman rowing me out to the island why on earth they bothered taking cows out to the island. It must be a great deal of trouble to load them on a barge and go back and forth. I have never forgotten his answer, "Shar, man, isn't the grass finer out there than on the mainland itself?"

D.W. Dunlop and Barbara Schmidt, in their chapter of the book *Biological Effects of Magnetic Fields* (edited by M.F. Barnothy), describe the effect of strong magnetic fields (1,000 G. and above) on plant tissue. The freshwater alga Pithophoro sp. was adversely affected by the strong field. Growth rates decreased and filaments became irregularly sinewy. They used a powerful horseshoe magnet so that the plants were in the field force of both magnetic poles.

Albert Roy Davis and Walter Rawls, Jr., in their fascinating little book *Magnetism and Its Effects on the Living System,* point out that the two poles of a magnet are of opposite polarity. In their experiments, they treated wheat seeds before planting. They obtained larger plants from seeds exposed to the south pole and smaller plants from seeds exposed to the north pole. When exposed to the south pole force-field, vegetables and some root plants such as

Radishes at top planted in check with no round tower in center of pot. Radishes at bottom were planted in a pot with a model carborundum round tower in the middle. Note more seedlings and a better root system on the bottom row of radishes.

sugar beets produced remarkable growth. Length and size of roots were greater and the cycle of growth was accelerated. Sugar beets yielded more sugar. The north pole gave the opposite effect.

Dr. Joe Nichols, in a fascinating interview in *Acres U.S.A.*, November 1980, discussed work in Canada with seeds and magnetism. The Canadian researcher found that there was much less starch breakdown in seeds of barley and winter wheat, when treated with magnetism. An Italian researcher found that magnetism inhibited seeds placed in a magnetic field with the radical end facing the south pole. It required only 100 G. for 10 days.

The great Russian scientist A.S. Presman, in his book *Electromagnetic Fields and Life* (translated by F.R.A. Brown, Jr.), quotes very elegant and detailed work by V.I. Karmilov (1948) showing that a magnetic field increases the yield of tomatoes. It also accelerates ripening according to A. Boe and D. Salunkhe (1963). A.V. Krylov and G.A. Tarakanova (1960) have shown that the root system of rye and beans increases and wheat and corn germinate much more rapidly in magnetic fields. Their graph of the increase in growth rate of the roots of barley seedlings is fantastic. It is indeed unfortunate that most of this kind of work is going on in Russia and not in the U.S.A. It does not pay for American scientists to ignore the Russian literature.

On a hike around the world over 30 years ago, observations of the form of religious structures were a clue to my concept of insect and plant spines as nature's antennae. I would, however, like to point out that most of the more elegant research on plants and magnetism has dealt with powerful, high-energy magnets and not, in any single instance, with the diamagnetic (repulsive) properties of the plants themselves, nor the paramagnetic (attractive) properties of the soil or clay flowerpots surrounding the plants.

Despite the fact that the magnetic field of the earth averages only about 1 G. and that there are paramagnetic forces in the soil around plant roots, no one has experimented on plants with the physicist's dia- and paramagnetic concept of low-energy magnetism.

It should be obvious to my reader by now that I am talking about the concept of "doping"—that is, as I like to say, "A little is a lot." I do not mean "doping" with trace elements or vitamins—which is also necessary—but rather "doping" with frequencies and magnetic fields. Doping is a fact of biological life and invariably its effects are long-term and difficult to detect. Could it not be that the good Irish monks were collecting a little of that cosmic paramagnetic energy and focusing it with round tower antennae onto the earth spots where they planted their crops—"doping" the plants with the energy? My round tower experiments tell me that is exactly what was happening!

I must add that there is considerable evidence that a weak magnetic field can increase one's memory. Why not? Magnetic tape recorders are memory machines. Remember the witch's hat and dunce hat of the ancients.

When I dream, I like to picture in my own mind the friendly Irish monks chanting away, surrounded by the paramagnetic energy "modes" focused at each floor of their heaven-pointed round towers. What earth-loving persons those Celtic friars must have been! God, no doubt, smiled just a little as He looked down on those monks and their beautiful, little, green island. Today, Ireland is rapidly moving to a petroleum-based, high-energy system. I wonder if He will continue to smile in such a manner at their modern folly?

4
SOIL ANTENNAE AND LIVING AURAS

"Today the country is merely the adjunct of the cities, and in these the balance of power entirely lies; the dwindling rural districts have been degraded into a mere food factory to supply the teeming millions of the towns. The more universally town rule and factory regulations are brought into the country, the more materialistic the country must become and the more divorced from the beauty and from the spirit of nature."
—**From Irish Earth Folk by Diarmuid MacManus**

I'll never forget my first view of the Great Pyramid. I came at it on the back of a camel led by Farag Suleiman, my desert dragoman. I still have dragoman Farag's card in my scrapbook from my walk in the sun around the world.

In 1948 I had never heard of a dragoman and when I looked up the word, I discovered it meant "an interpreter and guide for desert tenderfeet or tourists," so I must have qualified as the former. I met dragoman Farag in the famous Shepheard's Hotel in Cairo. His card read "Contractor for Sightseeing, Lower and Upper Egypt, Camping in the Desert, etc." I didn't consider myself a tenderfoot, but be that as it may, I had reached Cairo after a hard trek across the Middle East and was ready to "let somebody else do the driving."

The Shepheard's Hotel burned down shortly after I stayed there. The atmosphere of the old Shepheard's was right out of a William Somerset Maugham novel and dragoman Farag's Sidney Greenstreet shape added a mystique to my short camel ride out of all proportion to the actual fact.

There has probably been more baloney written about the Great Pyramid at Giza than any other structure in the world, but my dragoman embellished his pyramid stories with enough spice to

The author at Giza in December, 1948. Photo by Dragoman Farag Suleiman

convince me that somewhere within all the theories and speculation were some excellent clues as to why the structure was really built.

In my later readings on the pyramids of Egypt, I was surprised to learn that of the 30 or so different Egyptian pyramids, not a single mummy had ever been taken from inside a chamber. Lots of mummies have been found, but in most cases, they were buried in underground chambers around or under the pyramids and not inside of them. It makes about as much sense to say that a pyramid was a tomb as it would for the living, 1,000 years from now, to insist that every English village church is a tomb because it is surrounded by gravestones or that Westminster Abbey, with hundreds of bodies in its vaults, is a tomb. What the pyramids must

Soil Antennae and Living Auras 39

have been were "Egyptian cathedrals"—in short, places where the priests went to practice their Egyptian faith.

Even back in 1948 those structures, as far as I was concerned, fell in the same category as Gothic cathedrals, the Chedi of Thailand, round towers, pagoda, and other such strange and intriguing religious forms. As a matter of fact, the Great Pyramid at Giza is not only a stone pyramid but also a stone pagoda. The chamber inside is not a simple rectangular vault but rather a five-story, 80-foot-high stone tower. It is not at all unlike the floor design of certain Chinese stone pagodas.

We moderns tend to form an image of people such as the Egyptians as strange, way-out, mysterious people—a cruel people who beat slaves and forced them to labor long hours building the pyramids. There is no more evidence that unwilling slaves built the pyramids than there is that slaves built the Gothic cathedrals of Europe. As Kenneth Clark said in his beautiful essay, "The Uniqueness of Ancient Egypt," (in *Reader's Digest, Secrets of the Past*): "Egypt had a belief in the individual as a moral human being; in the beauty and dignity of man who had a soul that would survive after death; an awareness of nature as something beautiful as well as useful and very close to man himself; a well-organized system of government; and an art of unsurpassed grandeur."

Now I ask you, what is the difference between the belief and life of ancient Egypt and modern American Christian belief, except in our respect for the human body, we mummify it with formaldehyde instead of dry, desert air?

Strangely enough, our picture of God on every dollar bill is a pyramid symbolizing the Holy Trinity. The Egyptian symbol of God was Horus (the lofty one) depicted by a falcon. Mummified falcons by the thousands have been found in Egyptian tombs. In later dynasties Horus became identified with Re, the sun god, and the falcon is depicted with a sun disc over its head.

We will leave the finer interpretation of all this beautiful symbolism to the archaeologists and philosophers, however, I am certain of one thing—I have been a falconer since I was 14 years old and recognize the various species. In all of the Egyptian art, whether sculpture or drawing, the short-tailed, mid-eastern saker falcon (*Falco cherrig*) is not represented. This species is not even endemic to Egypt. What is shown is the longer-tailed kestrel (*Falco*

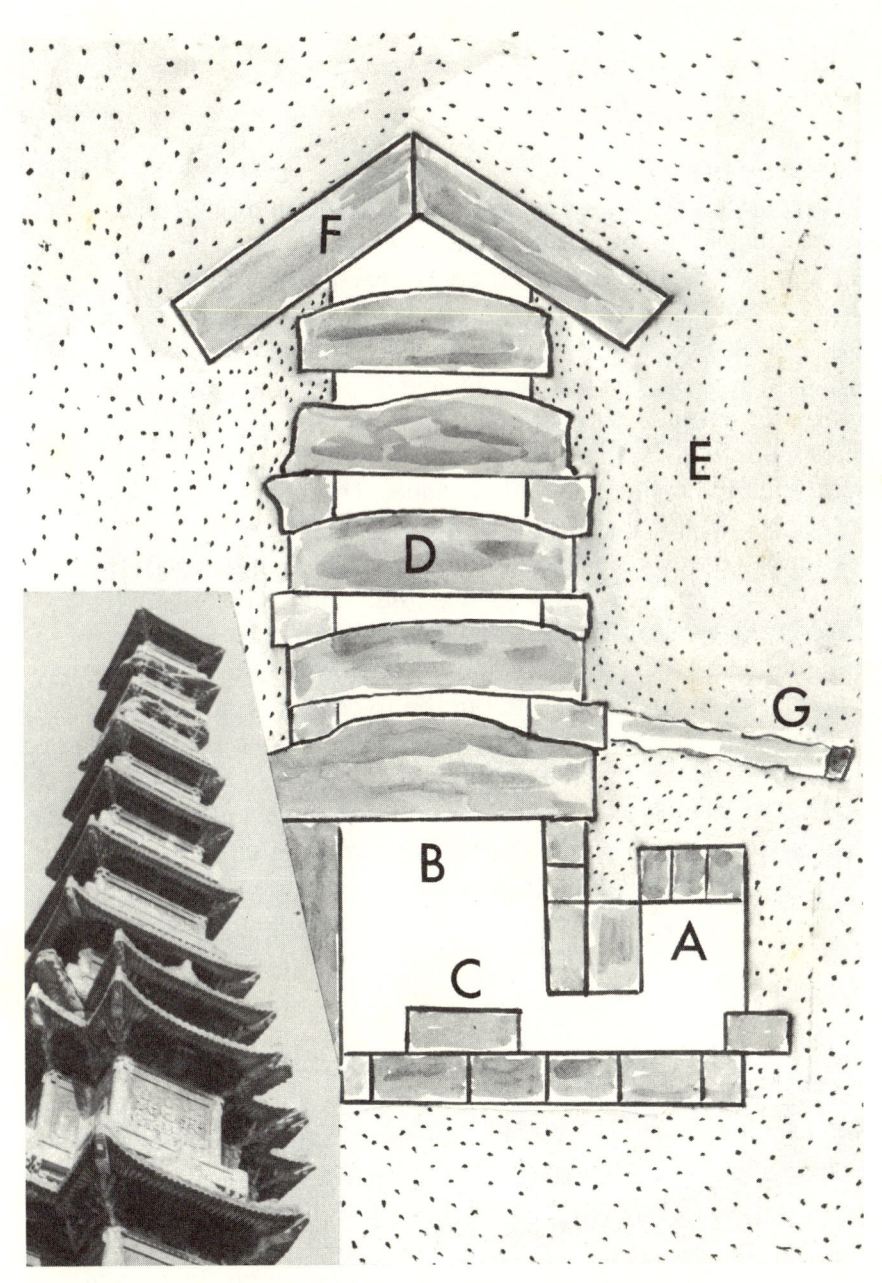

Pink granite tower inside the Great Pyramid at Giza in Egypt. It looks like a stone pagoda (see insert). A. entrance chamber; B. king's chamber; C. so-called sarcophagus; D. lens-like stone floors; E. pyramid proper; F. stone pagoda roof; G. workman's tunnel.

Soil Antennae and Living Auras   41

tinnunculus). The kestrel is the only bird known, other than the hummingbird and black-shouldered kite, that can hover suspended in midair as does the sun. The kestrel is endemic to Egypt and extremely common—which would not be true of migrating saker or peregrine falcons. Keep this kestrel symbol in mind as we continue this tale of stone and soil antennae.

Scattered around the world are numerous stone religious structures. Volumes have been written on the uses and meanings of such structures. They are usually found associated with monasteries where groups of religious live or places of worship such as cathedral steeples. The tapered Chedi of Thailand and Burma, for instance, are always found at wats (Buddhist monasteries) and are said to contain the relics of Buddha. There are literally tens of thousands of them scattered around southeast Asia.

In college and as a young entomologist, long before the scanning electron microscope was invented, I labored long and hard to describe the antenna sensilla (spines) on the corn earworm moth antenna. I had to make thousands of embedded wax slides and use a light microscope in order to put the microscopic shapes together in my mind. There are at least seven different shapes of sensilla on the corn earworm moth antenna alone!

As I drew my moth sensilla, I found myself saying, "Oh, there is a Siamese Chedi," or "What do you know—a church steeple or pagoda," or "That's a corrugated spine—like the Red Tower of India." And the last 25 years I have seen the shape of every religious structure I ever visited in my walk in the sun, and each in exact miniature on the antenna of an insect.

I realized many years ago that insect sensilla, which are micrometer in size, must resonate to micrometer-long electromagnetic waves, that is, infrared waves at room (life) temperatures. That should be obvious, but apparently it is not. Otherwise, more agricultural scientists would be studying such phenomenon. I also realized that the shapes of ancient religious structures must have some practical rationale other than the vague religious symbolism that we technocrats attribute to them in our ignorance.

Plants are diamagnetics—that is (I repeat myself for emphasis), plants have an inherent weak, fixed repelling force to a magnetic field. Michael Faraday, the ingenious English natural philosopher, discovered this in the mid-nineteenth century. His discovery has

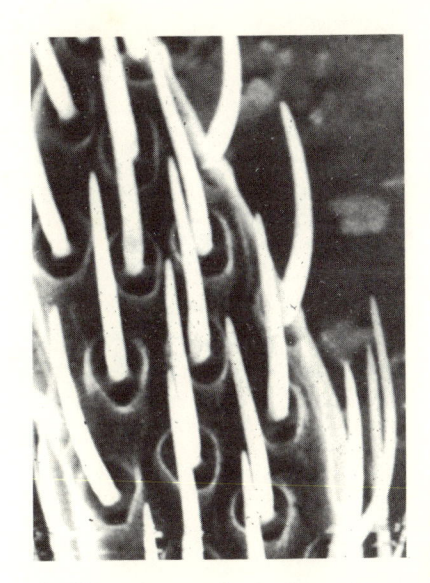

A Chinese pagoda in Hangchow, China. Hangchow, known as the "Heavenly Paradise" was considered by Marco Polo the most beautiful city in China. Insert—note that many diptera (flies) have pagoda-tapered sensilla on their antenna.

been ignored. John Tyndall, the great Irish natural philosopher, confirmed his friend's discovery by testing at least 30 different species of trees for diamagnetism. In the process, John Tyndall also became the world's first solid-state physicist. He noted that one of his sandpaper-smooth pieces of wood was weakly attracted to his magnetic field instead of repelled. When he examined it closely, he found that the wood had been branded with the small numbers "32." He realized immediately that his piece of oak was "doped" with the "essence of iron" and that "32" had a specific form (he used that very word, *form*, in italics.)

He considered that both the molecules of iron and the elongated form of the numbers changed the atoms of the wood so that the vibrating energies of the system were modified. His term was a new "mode of motion." He later wrote an entire book entitled *Heat as a*

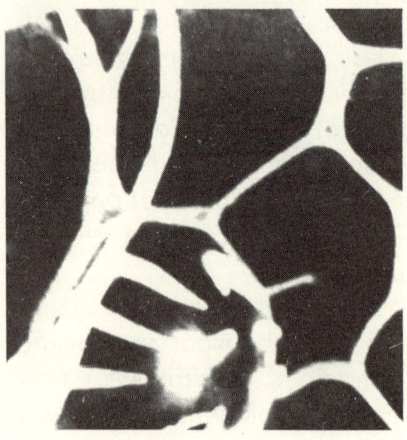

A Buddhist Chedi in Chang Mai, northern Thailand. The Buddhist Chedi is a South Asian form of a pagoda. They are located at Wats (Buddhist monasteries) and are said to contain relics. Insert—The peach tree borer has both single Chedi (top) and circled Chedi on the antenna.

44 Ancient Mysteries, Modern Visions

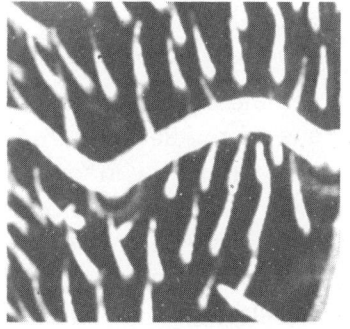

Stonehenge England is a huge circle of single and arched standing stones. Inserts—moths, such as the corn earworm, have corrugated "witches' rings" and flies, such as the Hessian fly which destroys wheat, have arched sensilla on their antenna.

Mode of Motion. In that book is not only the beginnings of all solid-state physics, with the concepts of "doping" and electron movement through solids "mode of motion," but also the basis for every bit of modern infrared technology. Bell Telephone likes to pretend that it discovered solid-state physics. This is, of course, ludicrous. Corporations and governments never discover anything. Individuals do!

Since religious structures of stone are doped with traces of various paramagnetic elements, as is the waxy exoskeleton of insects, it was only a short step for me to model different shapes and coat them with paramagnetic minerals or with beeswax. I deduced that if plants are diamagnetic, then animal matter, such as the human body, should be paramagnetic. It was also evident from my study of insect sensilla that species of insects detect animal matter—certain flies, mosquitoes, ants, etc., seemed to have a goodly number of corrugated sensilla on their antenna.

As I suspected, corrugated index-card round towers are considerably more sensitive to a magnet than are smooth-sided round towers.

New discoveries are made by individuals and not committees for the simple reason that the human mind is a computer storage bank of uncountable "life experience" files. A good scientist correlates and draws upon all such information no matter how trivial it might seem on the surface.

Somewhere in the dim memories of my past, back in 1948, I remember hiking across the flat plain of India with my friend Irwin Pless (now of Massachusetts Institute of Technology). We were tired, but when we reached the mysterious Qutb (pronounced Kutib) Minar, the Red Tower of Delhi, we decided to climb its steps to the balcony at the top. It was a strange ascent, for the expected fatigue of my already tired body seemed to absorb into the corrugated sandstone inner walls of the beautiful structure. It was as if the very weight of my body had lightened and new energy was infused into my being. Somewhat the same infusion occurs, but not as strongly, if one sits quietly in an Irish round tower.

The literature of holistic medicine and the occult is filled with works on a strange phenomenon called "animal magnetism." It is believed to be a factor in many so-called "occult" phenomenon. It is inherent to hypnotism, levitation and also divining or pendulum

Steeple on Norwich Cathedral in England and sensor on the cabbage looper antenna with its own steeple (cone) on it. The cone sensor is considered to be the antenna taste sensor on moths.

Soil Antennae and Living Auras 47

dowsing over the body (also called "radionics" in some holistic medical circles). It is known to depend upon certain special techniques such as vibrating and stroking. Insects do a lot of vibrating and stroking. If you don't believe that, watch a housefly sometime as it strokes it wings. Entomologists say they are cleaning their feet. That is about as silly as monks climbing into a tower to escape Vikings. If we could clean ourselves by rubbing our bodies, we wouldn't need Ivory soap. Rubbing creates static electricity and makes dirt particles stick faster than they would otherwise, and that is precisely the point to be made.

Rubbing creates a large electric field on living bodies that already have a fixed paramagnetic field and emits radiation in the far-infrared region of the spectrum. A human body with skin temperature of 96 F. centers at a wavelength of approximately 10 micrometers in the far infrared.

Now we see what is meant by the holistic viewpoint as applied to farming or medicine. We may separate out these forces by experimentation to study them, but the electric field, the paramagnetic field, and the electromagnetic field (infrared) do not operate separately in nature. Rather, they work together holistically to create a force called—by those that understand the holistic viewpoint—animal magnetism.

What does such a force mean in terms of such huge structures as the pyramid, or even the smaller monastic structures such as Chedi and round towers?

The parapsychological term for the ability to move an object without contact is psychokinesis or PK power. Only a few so-called PK mediums have such power. Apparently, they have trained themselves to become, so to speak, animal-magnetism antennae and thus project the energy much more forcefully than does the average person.

Some individuals, no doubt, have a suitable form (bodily structure or molecular makeup) to concentrate the energy.

If you follow my directions for making a corrugated round tower, then you will soon discover that everyone has the power of psychokiness inherent in their own body. (See Appendix).

How did I discover it, and why does it work? Quite simply, because I took the exact opposite viewpoint of other researchers working in the field of PK phenomenon—insight I believe it is call-

48 *Ancient Mysteries, Modern Visions*

A carborundum, paramagnetic round tower in the center of a radish planting. This tower was teardrop-shaped with sharp edge facing the reader. Good soils tested by the author were always highly paramagnetic, poor soils the opposite. There are some indications from these radish planting experiments that the paramagnetic energy can be controlled by the shape of the tower. There are so many variables that these experiments are preliminary, but note that for 180 degrees on the back rounded side, the six-day-old seedlings are up and strong, while for 180 degrees of the knife-edged side, they are stunted or have not emerged at all. In one experiment where I planted inside a stone circle there emerged a pattern where the north half sprouted and the south half did not, forming the Yin Yang pattern seen on the Korean flag. I also have noticed that around a concrete silo that is never used near Gainesville, Florida, weeds are twice the size of weeds in an adjoining lot on the same type of soil. The silo concrete is highly paramagnetic.

ed. Instead of concentrating on amplifying the energy from the human body, I designed a sensitive receiver based upon the form of insect sensilla and stone religious structures. The receiver (antenna detector) is tuned or matched to the infrared paramagnetic aura of the body. Instead of moving any old object on a table (a paper clip, spool, etc.), the aura is moving an object that, as antenna engineers say, completely absorbs the aura energy. Because the friction contained in a twisted thread is very low, the

Soil Antennae and Living Auras   49

energy "collected" by the corrugated round tower receiver easily overcomes the drag of the thread.

What does my experimental aura detector mean in terms of the various religious structures and the Great Pyramid of Egypt? In almost all cases (Chedi, round towers, etc.), the land around the structures is a veritable garden of paradise. It is my firm belief that the infrared paramagnetic forces radiate from the towers in waves from their base and increase the paramagnetic properties of the surrounding soil. I already have some evidence for this theory. It is rather easy to test soil by coating the towers with different soil types. Good soil, with the proper chemical-organic-solid-state makeup will be highly paramagnetic. Poor soil, such as low-organic, sandy types, will barely respond to even the strongest magnet.

It is but a short step from psychokinesis (moving a body sidewise) to levitation (moving a body upward). I am firmly convinced that the pyramids, both Egyptian and Central American, were huge antigravity structures for levitating the priests. The pyramid acted like a huge "enlarger-type" condensing lens which concentrated the cosmic energy into the hollow resonant stone tower, which in turn was filled with the IR-paramagnetic, organic breath of chanting priests. If my belief seems farfetched, then perhaps the "picture" language of the Egyptians themselves should give one cause to reconsider. There are quite a few Egyptian wall reliefs that show a Pharaoh lying on a sphinx-like stone couch. In the next accompanying relief, the carving shows the Pharaoh six inches above the surface of the couch with a sacred kestrel hovering above his outstretched body. Remember the kestrel is one of the few large birds that can hover in a fixed position in space.

The pre-Aztec Pyramid of the Sun outside Mexico City is built from volcanic rock and is highly paramagnetic. I have tested it. The top was believed to be used for human sacrifice and was thus heavily doped with blood and organic human effluvia. The Aztec language is Nahuatl and the Nahautl word for the pyramid is "Teotihuacan," meaning "place where men become gods." What better way to become a priestly god to your followers than to rise ever so gently into the air!

Colin Wilson, in his excellent book on the occult entitled

Mysteries (in a chapter on the rediscovery of magic) states, "The anomaly vanishes if we can reconcile ourselves to the idea that magic is not a branch of the supernatural, but an acquired skill, like repairing cars or performing on a trapeze."

I leave you with the thought that my round tower aura sensors are neither magic nor supernatural, but rather they are an elegant and inexpensive way to gain insight into the workings of the soil and the human body. Both are subjects in need of constant study, lest we destroy both the land and our beings. That, of course, is what eco-agriculture seeks to prevent by celebrating both the beauty and the spirit of nature.

52 Ancient Mysteries, Modern Visions

5
STONE AND CLAY —
THE REAL SECRET OF THE PHARAOHS

"Accursed be the soil because of you. With suffering shall you get your food from it every day of your life. It shall yield you brambles and thistles, and you shall eat wild plants. With sweat on your brow shall you eat your bread, until you return to the soil, as you were taken from it. For dust you are and to dust you shall return."—**Genesis (3)17-19, The Jerusalem Bible.**

Life has a strange way of causing one to pause and consider that inexplicable word "fate." In another chapter, I discussed my belief that the Great Pyramid at Giza was a huge stone cathedral where the priests and pharaohs went to perform the mystic ceremonies of their ancient agricultural civilization. My thesis is that the pyramid is a huge paramagnetic antenna capable of "focusing" like a lens waves of paramagnetism into a tower like the King's Chamber.

I received a letter from Dr. Abdel Gawaad, secretary of the First International Congress on Soil Pollution. Professor Gawaad is an entomologist at Zagazig Agricultural University in Zagazig, Egypt. The Egyptian government, the letter read, would pay my way to Cairo if I would lecture on my research concerning the control of insects and plant growth with infrared and paramagnetic wavelengths. Since the letter had taken 40 days to reach the United States, I cabled my acceptance. My wife and I were soon on our way.

The agriculture of the Nile Valley is in serious trouble. The Egyptians grow three crops a year of fine, long-fiber cotton. Three crops means that they dump three times as much insecticide into their Nile soil as we do in our own one-crop South.

Even more tragic, where agricultural production is concerned, was the construction of the High Dam at Aswan in upper Egypt. The Nile River is a long narrow oasis sandwiched between two vast, deadly deserts—the Libyan or Western Desert and the Eastern Sahara esh Sharqiya or Arabian Desert. I call them "deadly deserts" after the desert in one of those childhood masterpieces, *Ozma of Oz*. To be lost in one or the other of them is to perish.

Previously there was a low dam at Aswan. It allowed that magic paramagnetic clay from deep in Africa to flow in suspension over the dam and revitalized the sandy banks of the Nile. With the advent of the High Dam, the magic clay no longer flows during the June inundation. Rather, it silts up in the lake behind the huge structure.

In ancient times, the priests of the temple developed a warning mechanism termed, in modern jargon, a "Nileometer." When the water reached a certain level in the temple Nileometer, the priests sent forth a warning for the farmers to move to higher ground.

As J.E. Manship White said in his elegant little book, *Ancient Egypt*, "The human story of that splendid civilization must be unfolded against the natural background of river and rock, sky and sand. Any study of it must be prefaced with a brief outline of the environmental factors involved."

It's too bad modern economists do not study such an outline. What is unfolding today is a 6,000-year-old fertile land succumbing to the false promise of Western industrialization; a land that has few of its own natural resources to support such an urbanized system. Cairo, which I knew thirty years ago as a beautiful agricultural town center, is today the most air- and soil-polluted megalopolis (with about 14,000,000 people) I have traveled to in the last fifteen years. In short, after 6,000 years of organic abundance, the Nile Valley, if not dead, is at very best slowly dying. Egypt, which used to export food, now imports almost half its daily sustenance. As Professor White has written, "Egypt is the Nile." Alas, today the Nile is being murdered.

There is nothing new, of course, about ancient man's knowledge of the "magic force" in clay and stone. The Egyptians knew all about the medical and agricultural importance of their sacred stone and clay. They might not have called the force "paramagnetism," which is a modern word, but they knew how to design stone antennae to collect and amplify the force. Archaeologists' studies show that as early as 3300 B.C. the Egyptians fashioned superb stone tools and weapons. Their mud brick houses and grain silos reached a high degree of sophistication.

The stone pyramids are believed to have been built about 2600 B.C. during the period of the Old Kingdom, 2778 to 2300 B.C. (4th Dynasty).

As late as the time of Christ, the curative power of clay was known. In Book One of the "Essene Gospel of Peace" (translated by E.B. Szekely) we read:

"And there were many sick among them tormented with grievous pains, and they hardly crawled to Jesus' feet. For they could no longer walk upon their feet. They said, 'Master, we are grievously tormented with pain; tell us what we shall do.' And they showed Jesus their feet in which the bones were twisted and knotted and said, 'Neither the angel of air, nor of water, nor of sunshine has assuaged our pains, notwithstanding that we baptized ourselves and do fast and pray, and follow your words in all things.'

" 'I tell you truly, your bones will be healed. Be not discouraged, but seek for cure nigh the healer of bones, the angel of earth. For thence were your bones taken, and thither will they return.'

"And he pointed with his hand to where the running of the water and sun's heat had softened to clayey mud the earth by the edge of the water. 'Sink your feet in the mire, that the embrace of the angel of earth may draw out from your bones all uncleanness and all disease. And you will see Satan and your pains fly from the embrace of the angel of earth. And the knots of your bones will vanish away, and they will be straightened and all your pains will disappear.'

"And the sick followed his words, for they knew that they would be healed."

The summer before we went to Egypt, my wife developed a cyst on the top of her instep. It might make hiking in Egypt very difficult, so the doctor advised surgery. I ground up some volcanic rock and clay from an old flowerpot. She wore it over her instep in a cloth bandage for a week and the cyst disappeared. Of course, my degree is Ph.D., not M.D., so I certainly did not "cure" her. God did that through the intercession of paramagnetic stone and clay.

As I have pointed out, root growth can be controlled by paramagnetism. Good soil is always highly paramagnetic. Some very

Stone and Clay

From the top of Jade Mountain I could see the fields planted alternating wedges radiating from a central hub like a wheel. It was the Pah Kua or lucky symbol of agriculture. The Pah Kua is near Hangchow, China.

preliminary experiments indicated that such energy can indeed be treated as a wave phenomena. It travels like ripples of water out from a model round tower or Egyptian obelisk to where it is absorbed if plants are seeded in a circle around the base of the tower. There is nothing new about energy in planted circles either. Derald G. Langham has written an entire book on the effectiveness of planting in circles entitled *Circle Gardening*. The Chinese have planted in circles for thousands of years. The Pah Kua near Hangchow, China is considered a lucky symbol of agriculture. There is considerable evidence that lush, beautiful gardens surrounded the pyramids and tall, square, granite obelisks of Egypt. The famed hanging garden of Babylon grew from terraces on the slopes of a step pyramid. All of the many descriptions of these early stone structures indicated to me the possibility that they were knowingly designed as giant paramagnetic antennae. If so, then the Great Pyramid at Giza must, from its great mass alone, be one of the most efficient of such energy collectors.

Ever since my trek around the world in 1948, I have been fascinated by the shape of stone religious structures. In that walk in the sun, I passed by Buddhist Chedi pagodas, Irish round towers, stone rings and megaliths, obelisks and, of course, the pyramids of Egypt.

Among the large pyramids, the Great Pyramid is by far the most interesting. Both from the standpoint of sheer mass and also from the strange, stone, pagoda-like structure inside called the King's Chamber. The fact that no mummy has ever been taken out of any of the pyramids nor the walls decorated with colorful hieroglyphs as are Egyptian tombs does not seem to curb the foolish notion that pyramids of Egypt were the tombs of pharaohs. Even if mummies were found inside, it would not necessarily be their primary function. As I have said, one then might as well consider the great Gothic cathedrals or St. Peter's as tombs since all of them contain many graves of famous persons.

In Egypt, bodies have been taken from under or around pyramids, but never from inside the differently shaped inner chambers.

We have all heard of the sacred scarab beetle of Egypt. What is much less well known is the importance of the vespid wasps in Egyptian lore. Along with the scarab beetle, the vespid is the most common hieroglyphic insect carving found on temple walls. Is it coincidence that the antenna of the paper wasp vespid has both pyramids and corrugated round towers on it? The family Vespidae includes mud daubers and paper wasps. As you will recall, the corrugated round tower is one of the most sensitive paramagnetic-infrared, waveguide, dielectric resonators. The tiny structures, called sensilla by entomologists, are exact duplicates, with a wax coating, of round towers, Chedi, pyramids, etc. There are even Stonehenge loops on some species of diptera (flies). These peculiar-shaped spines are the resonators to the coherent infrared energy given off by the scent molecules that attract insects to their mate or host plant.

I have long considered the stone religious structures of the world to be dielectric waveguide resonators of paramagnetic energy. In other words, as in the case of infrared or visible radiation, the force called paramagnetism actually travels as waves. Therefore conventional dielectric (silicon is dielectric) waveguide theory applies to

The antenna sensilla of a wasp showing pyramidal sensilla and corrugated sensilla—two of the best configurations for focusing and concentrating the paramagnetic force. The species, Polistes metricus, *is a vespid wasp. Although seldom mentioned in the popular literature, the vespid hieroglyph is even more common in Egyptian lore than the sacred scarab beetle.*

the great stone structures just as it does to optical and radio antenna design—and to my little wax-coated insect sensilla.

What exactly do I mean by "dielectric paramagnetic antenna"? A dielectric substance is an insulating material as distinguished from a conducting metal. The pyramids are built mainly of huge limestone blocks. Limestone is largely calcite with dolomite and traces of quartz and silicates. None of these substances are good conductors of electricity. However, most limestone also contains clay minerals from chemical or biochemical precipitation during its formation. Limestone, therefore, contains traces of sulfides, iron and manganese hydroxides. Since the stone is "doped" with traces of iron and manganese, it would be closer to the truth to call limestone a paramagnetic semiconductor. In other words, under the proper conditions, it will not only conduct electricity, but also amplify cosmic paramagnetic forces.

Paramagnetism, as described in the nineteenth century when it was discovered, is a weak, fixed susceptibility toward a magnet. The key words are weak and fixed. Weak is self-explanatory. By fixed, physicists mean that it is "inherent" in the substance and cannot be transferred as one does by rubbing a nail or screwdriver against a strong magnet—the rubbed metal becomes magnetic. Of course, most physicists do not consider paramagnetism as having a wave form comparable to electromagnetic energy as I do.

I have been working with paramagnetism for years. I not only discovered good soil to be paramagnetic, but likewise refined beeswax. That later discovery indicated to me that since insects are coated with wax, their antennae are paramagnetic structures. The next bit of reasoning followed from my insect work and it involved asking the right questions about religious structures such as pyramids. The right question is, "Since insect sensilla have definite shapes for resonating to different infrared wavelengths, do the various shapes of stone religious strucures indicate that they resonate to the paramagnetic forces of the cosmos? In other words, can they collect and concentrate the paramagnetic cosmic force above what would normally occur in unshaped soil and rocks? The answer is yes, and inside that King's Chamber at Giza I proved that without a shadow of doubt.

In my work with Irish round towers, I developed model smooth and corrugated round towers that respond to the paramagnetic-infrared aura of the human body.

Pink granite is the most paramagnetic stone that I have ever measured—with an instrument called a magnaprobe. The inside tower chamber of the Great Pyramid is constructed from beautiful pink granite. Each of the giant pink granite lenses (tower floors) act to focus the cosmic paramagnetism down (like light waves) into the King's Chamber. The outer limestone portion of the pyramid serves as a giant condenser lens—as in a photographic enlarger—to diffuse and concentrate paramagnetic waves. Of course, the outer, smooth, tura limestone casing of the Great Pyramid has been destroyed so it will never again stimulate total levitation of a heavy body.

My round tower sensors will, of course, move sideways to a human body (telekinesis) anywhere on the earth. The arc of movement in my house is 60-70 degrees, but in the King's Chamber it

Stone and Clay

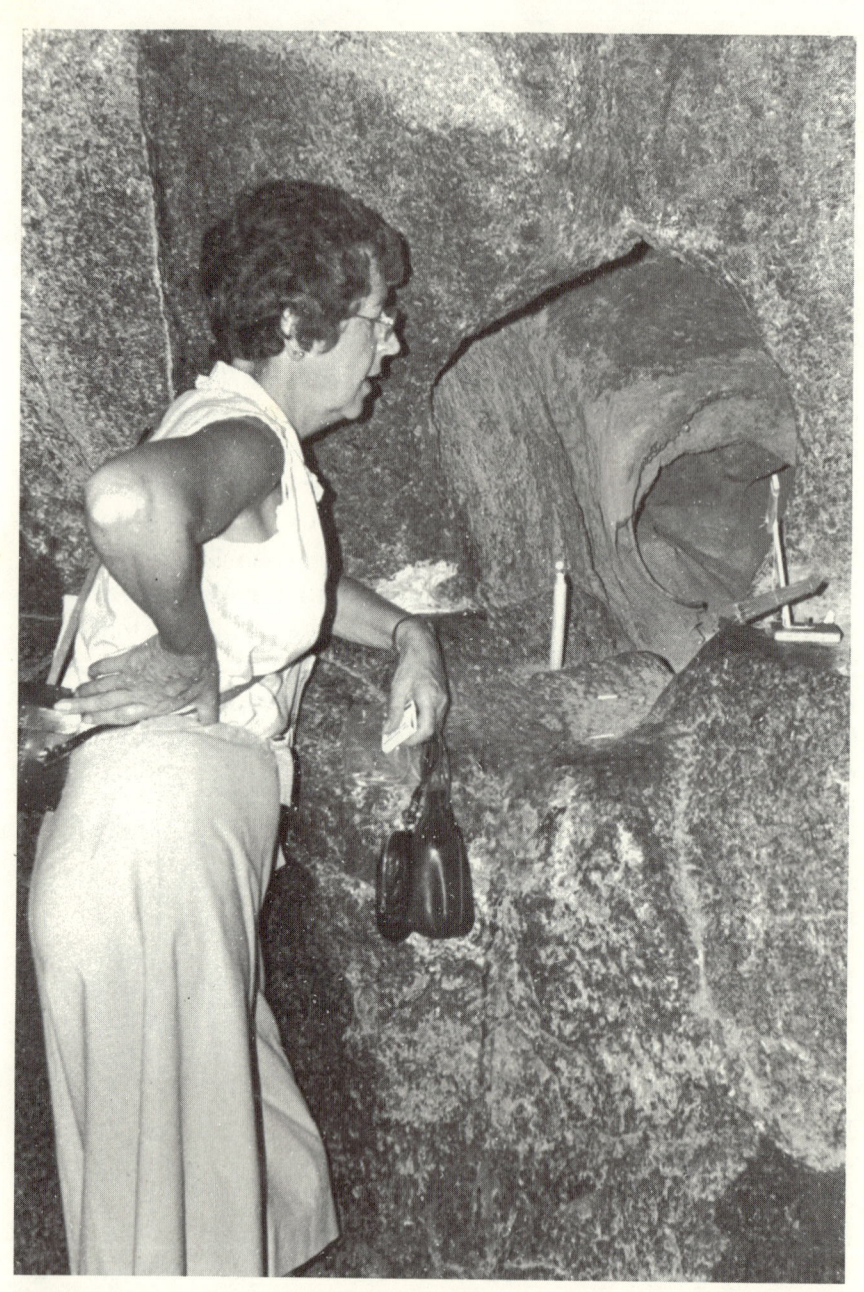

The author's wife observing the round tower aura sensor at the south air vent in the King's Chamber of the Great Pyramid. Note that at the airvent the light candle does not flicker. The sensor not only oriented itself to the human body over 300 degrees of arc (telekinesis), but also rocked violently up and down (levitation) every time a human body approached it.

moved 200-300 degrees in a steady sweep to the human aura. The sensor was five to ten times as sensitive inside the chamber as inside my house in Gainesville, Florida.

My wife and I were in the pyramid chamber for two hours by ourselves. But later, as tourists began to collect inside, we utilized German, French and English bodies in our telekinesis experiments. I had to give more than one spontaneous lecture on the phenomenon of paramagnetism to those who helped us. We can state, in regard to the force from the paramagnetic human, that there is no such thing as a national boundary. All living creatures are paramagnetic and the "doping" of air in the King's Chamber with living organic molecules is a necessary part of the levitating system.

The historical evidence for priestly levitation inside the pyramid chambers of Egypt is the subject of a long literature search by myself. Suffice it to say that Horus, the sun god of the Egyptians, is modeled after the little kestrel falcon and the kestrel is one of the very few birds that can hover stationary in midair—suspended, so to speak, between heaven and earth.

On the dust jacket of Peter Tompkins' wonderful book, *Secrets of the Great Pyramid*—and it misses the real secret—there is reproduced a beautiful series of ancient stone-wall reliefs. These wall reliefs show an Egyptian priestess raising her hands above the outstretched body of a pharaoh lying on a typical Egyptian couch. In the next panel, the pharaoh is six inches off the couch and above him hovers a strange bird. I say "strange" because the bird has the head and body of a falcon, but with much broader, black-shouldered wings, so I could not recognize it. Although the pharaoh is obviously levitating, the mixed figure of the hybrid bird of prey has puzzled me for years—that is, until fate stepped in to clarify the hovering symbol.

August is hot in the desert, so the tourist season is in wintertime. For that reason, there were few tour boats on the Nile River. Again, fate stepped in with the solution. We met a professor, with the rather appropriate name of Ramisi, and he arranged to get us on the river boat named the Karnak. Karnak is the great temple at Luxor. That temple makes the Parthenon of Greece look like second-class work.

The Karnak had been assigned to a group of 30 Danish devo-

The American Kestrel, Falco sparverius, *like its European and African cousin the common Kestrel,* Falco tinnunculus, *has a "football helmet" pattern on its head. Like its African cousin it is also a master at hovering. The American kestrel is as common at Hueco Tanks as the common kestrel is at Egyptian temples and pyramids.*

tees of Egyptology and their archaeological guide. We were attached to the group with our own private English-speaking guide. The riverboat would stop at all the great temple sites of the Upper Nile between Luxor and Aswan. We flew to Luxor to join the group.

Before leaving for Egypt, some of my ornithologically slanted friends, and I must include myself, thought that perhaps due to the great amount of DDT in the soil, there would be few kestrels left in Egypt. DDT thins the eggshells of hawks and falcons. We were totally wrong. During our five-week stay in Egypt, I counted 72 kestrels, almost all roosting on the minarets of Moslem mosques, ancient temple walls, and even on the sides of the Great Pyramid itself. The lesser kestrel of Egypt is apparently social and roosts in groups. Since all were seen in cities and villages, the species has obviously become an urban and temple wall dweller and hunts its food supply from city mice.

Never in my wildest dreams did I suspect that one little falcon would, so to speak, fall into my very own hand at an appropriate holy spot, and thus accentuate my entire thesis concerning the Great Pyramid. It happened at the river town of Edfu where we stopped to visit the temple to Horus the falcon god.

The author's wife sits in front of the pink granite statue of Horus, the falcon god of Egypt. It is at the temple of Edfu, sixty miles north of Aswan in Upper Egypt.

Stone and Clay 63

I had just come from the temple sanctuary when my wife said she heard a kestrel. I scanned high up on the temple walls but it seemed to be calling from behind me. I soon spotted it on the fist of a temple guard not fifty feet from the magnificent pink granite statue of Horus at the temple entrance. The guard obligingly perched the little bird on my fist. The kestrel, which had fallen from its temple wall nest in June, drew itself up in the exact position of Horus, the granite falcon god. The match was startling; an out-thrust breast, the sweep of the long, curved tail, and the extremely long, slender legs of the kestrel god.

It was at Kom Ombo the next day that I received an answer to the hybrid falcon of the Egyptian wall reliefs. High above the Nile in front of the Crocodile Temple at Kom Ombo soared a beautiful white and black-shouldered kite. It drifted above the temple walls and suddenly stopped in mid-air. Barely moving its wings, its legs hanging downward as in the reliefs, it "parked" directly over my head. It hovered like the white pure soul of a nature god between heaven and earth. The black shoulder feathers and the black ring around its red-colored eye was visible from the top of the temple wall where I was standing. The kite's eye is the exact duplicate of the red-colored, black-circled, sacred eye of Egypt as seen on the pigmented stone walls of certain tombs in the Valley of the Kings at Luxor.

Suddenly, at that instant in time, it all came together in my mind—the great stone antennae for cosmic energy, the paramagnetic soil and clay, a gift from the gods, and finally the mature love of those man-gods, the pharaohs. Deep in the Great Pyramid they breathed out that mixture we call breath—but which the Orientals call "the spirit of life,"—and chanted to modulate the potent vapor which was then energy magnified by the great stone, paramagnetic pyramid. Ever so gently they rose in the air just as we Catholics raise the bread—the body and blood of Christ—to the heavens above at our own Mass. Those learned pharaohs and priests of ancient Egypt elevated their own bodies and suspended it, so to speak, between heaven and earth. Their very own wall pictographs, the physics of their great pyramids, and lastly the symbolism of their falcon-kite god tell us that this is so.

On the last day of our river journey, we docked all afternoon by the temple at Kom Ombo. That night, as the start of the great

The author and his guide with a lesser kestrel. Note the long legs and tail—the exact form of the Egyptian Horus statues. The kestrel can hover in mid-air and is commonly seen nesting on the sides of pyramids and temples.

vibrating diesel engines awakened me at 3 a.m., I arose and went up to the top deck. It was silent and deserted in the dim light of a false dawn. The Big Dipper and the North Star were in heaven, and as rays of sun broke the horizon, the melancholy prayers from bankside village mosques floated across the river.

My mind drifted back to Mideastern desert nights over thirty years ago. I was a far different body then than now. In thirty years, every cell in my body has been replaced six times. But then, that is only a worldly body for my soul has not changed one tiny bit. The ancient agricultural Egyptians understand this better than modern man does. As I lay back on that gentle, moving cradle of the Nile and as the energy from the river, rock, sky, and sand caressed my body, I knew why I was at that moment in time exactly where I belonged. I also knew why I had become an agricultural scientist.

My father, who fought in Burma with the Chinese in World War II, was a remarkable person. Deep in the jungles he once carved a little wooden figure. When he returned home after the war, he gave it to me and about it I once wrote:

> *it's shaped like real, this wooden beggar,*
> *that sits crouched on my desk,*
> *carved from mahogany wood.*
>
> *its eyes are drilled deep,*
> *two holes in the brown wood*
> *and between its perching bird bent feet*
> *a begging bowl for food or money, or whatever*
> *a beggar needs to eat.*
>
> *I've passed his kind a thousand times*
> *in the misery of guttered streets.*
>
> *I've walked with my camera 'round my neck and*
> *jingling coins to spend for film or books about*
> *starving beggars in far off places*
>
> *as I passed I wondered why he is he*
> *and I am I*
> *Shall I ask God?*

66 Ancient Mysteries, Modern Visions

Not a single agricultural scientist has ever received a Nobel prize for science, and I doubt if any of the "Einsteins" of agriculture ever will. I also doubt if any really care. I have long believed that if my work should aid one single farmer in growing one more measure of the bread of life, and if that loaf should nourish even one more living soul by filling a single wooden bowl, then the meaning of my soul shall be fulfilled and the voice of those forgotten peoples shall have touched my heart.

6
A New Look At Stone

Down through the ages stones have been used as healing agents and sacred charms. The index of the book by Sir James George Fraizer, entitled *The Golden Bough—A Study in Magic and Religion,* lists quite a number of ceremonial rites where stones are utilized for healing and where fertility of the soil is the main concern. Sir James Fraizer, of course, treats these so-called fertility ceremonies as the superstitious ravings of a backward people. This is the usual condescending attitude of the modern, high-energy, inorganic researcher toward our low-energy, organic ancestors.

The ancients, according to *The Golden Bough,* used stone in their ceremonies to facilitate childbirth, to cure jaundice, to use as a homeopathic remedy, to make rain or as a fertility charm, as sunshine and wind charms, and most interestingly as a charm for fatigue transferal.

People of the Babar Archipelago rub and strike themselves with stones believing (as Sir Frazier puts it) that they transfer their own fatigue to the stone. They throw the stone in places which are set apart for that purpose. It is the practice that has apparently given rise to the rock cairns (piles) that are found beside paths and on mountaintops all over the world.

You will recall how after a long desert walk in India, I climbed the beautiful Red Tower of Delhi (Qutb Minar) and the fatigue of my body dissipated. I arrived at the bottom fresher than when I started the 280-foot climb up the steep stairs.

I do not think that the ancients merely believed that they would transfer fatigue to the stone. I think that they knew they could—for it has also happened to me hundreds of times during my life other than at the Red Tower. I have been rejuvenated over and over because in my youth I became a falconer, and falcons nest on rock cliffs. To get a falcon, one has to either trap the bird or scale a steep rock cliff to the eyrie.

Rock climbers often speak about the *climber's high*. The great British rock climber Joe Brown, in his book, *The Hard Years,* has refuted the idea that climbers have a phobia for death and that fear is the stimulant for their sport. He wrote, "Climbing is probably one of the greatest of all emotional stimulants, and without emotions, man may as well be a vegetable. To a non-climber, the obvious emotion triggered off by climbing is fear! If this were true, climbing would never have become the popular pastime it is today."

When George Leigh-Mallory, who lost his life on Mount Everest, was asked why he wanted to climb the mountain, all he could think to answer was "because it is there." That, of course, is a non-answer and demonstrated that climbers don't even understand the power in rocks.

There are books on the *Joy of Sex,* the *Joy of Cooking,* etc. I call rock climbing the *Joy of Clinging* and it has nothing to do with the stimulation of fear. At any rate, most climbers aren't afraid or they would not be climbing. The climber's high, rather, has to do with the effect of paramagnetic rock on our own paramagnetic body. Paramagnetism is a weak susceptibility to a magnetic force. In other words, certain kinds of stones are weakly attracted to a strong magnet. Climbers, so to speak, absorb the rock energy into their own bodies. Not only does that overcome fatigue, but it also puts *joy* into the brain.

What I call *joy in the brain* is a form of tranquilizing—but not in the sense of what happens when one takes a tranquilizing drug. Tranquilizers, like wine or beer, make one happy, but the drug also *dopes* the body and disturbs the normal functioning of the brain. Reaction times are slowed and thought processes made less efficient. No rock climber in his right mind would take drugs or alcohol before starting a hard climb.

Climbing a cliff or meditating in a stone building stimulates joy (tranquilized state) without making mental mush out of one's brain. Indeed, my experiences in meditating in round towers, the King's Chamber of the Great Pyramid, and clinging to rock cliffs is that the thought processes are stimulated to higher levels of rationality—the exact opposite of the drug effect!

Of course, the ancients were wrong in one respect in that they weren't actually transferring their fatigue to a rock, but rather transferring rock energy to their tired bodies and thus overcoming fatigue. Instead of taking fatigue away, as such, they were taking it away by raising their own energy level, but the rock does not assume the fatigue in place of the human. Indeed, all of my experiments so far indicate that my sandpaper and rock round towers are actually paramagnetic solar antennae that collect the subtle magnetic radiation from the sun and pass it on to the plants. To understand what I mean, it is only necessary to set up a few simple experiments in pots, as illustrated on page 49.

I use radishes as my experimental subjects. I grow them in one-foot shallow, plastic pots under growlight fluorescent bulbs. That way the light is the same from all directions. You will note that all the plants bow towards the model tower in the center. One might assume that the white light reflecting from the tower was responsible for this attraction phenomenon. This is not so because a black carborundum round tower, which is even more paramagnetic than white limestone, causes a greater attraction toward itself.

It is interesting that the seedlings always sprout quickest (one day lead) on the north side of the tower and do not grow as fast or as tall toward the south. The radishes that are planted due south of the tower are usually stunted and only half the size of the rest of the circling plants.

A quantitative measure of paramagnetism in rock or clay particles (ground-up flowerpot) can be calculated by weighing the amount of granules attracted to a magnet (susceptible) versus grains not attracted (non-susceptible). It will be noted that while clay flowerpots, granite, basalt and schist are all highly paramagnetic, limestone is so weak it can only be demonstrated to be paramagnetic by shaping a chunk into a round tower and hanging it on a thread (see table page 72). It will then swing its cone-shaped end toward the magnet. It is this latter shape phenomenon that indicates that the paramagnetic force goes out in waves and can thus be studied using waveguide (radio) design (Maxwell's equations) mathematics. If paramagnetism were not a wave function, then form would not be important. Just as certain forms are necessary for strong resonance in radio antennae, the same is true of stone paramagnetic antennae.

Another irrefutable proof that shape is important is the single fact that if you take a portion of ground-up rock or clay and regrind it, the ratio of susceptible to nonsusceptible will change. In one experiment, we see that we had 7.4 grams of homogeneous flowerpot and that 2.1 grams came to the magnet while 5.3 grams did not (non-suseptible). That, of course, gives a ratio of 1 (suseptible) to 2.5 (non-susceptible). I next took 5.77 grams from the 7.4 and hit it a few more times with the crushing hammer. I was then able to pick up 1.99 grams (susceptible) and 3.78 grams stayed behind (non-susceptible). This gives a ratio of 1 (susceptible) to 1.9 (non-susceptible)—a lower ratio than before. In other words, there is a statistical range of variation based on the pure chance that each time one crushes the sample more or less, particles will have the right shape for resonance, or the wrong shape. It is the shape that counts. Size and weight do not matter other than using a stronger magnet would lift more of the flowerpot pieces, larger and smaller.

I use a very strong, rare-earth magnet made of samarium cobalt. It is called a rare-earth magnet, and although only one inch round, it pulls 2,000 gauss. A mighty-mite of a magnet!

We can see from these ground-up flowerpot experiments that paramagnetism must definitely be a waveform susceptible to magnetism. If I take my same 2,000-gauss magnet and sprinkle the table with nuts, bolts, paper clips, iron filings, etc., of every diverse shape, the magnet will attract every last piece of the different metal forms, but you can never pick up all of the flowerpot granules no matter how large or small the grains are. Magnetic substances such as iron are 100% attracted to a magnet; paramagnetic substances such as stone and clay demonstrate ratios of attraction (susceptible) to non-attraction (non-susceptible) to the magnet. Once again, I must emphasize the attraction depends on shape.

Obviously if my paramagnetic-stone round towers are resonant wave receivers (susceptible) of magnetic energy, then the energy

RELATIVE PARAMAGNETIC SUSCEPTIBILITY VS. NONSUSCEPTIBILITY IN STONE AND CLAY MINERALS.
GAINESVILLE, FLORIDA 1982

Mineral	Weight in Grams	Susceptibility (gm)	Nonsusceptibility (gm)	Ratio[1]
Red clay flowerpot	7.40	2.1	5.3	1 to 2.5
Pink granite	16.1	5.8	10.3	1 to 1.78
Basalt	9.03	9.03	0	1 to 0
Garnet schist	24.3	22.0	2.21	10 to 1
Olite limestone[2]		0	0	0

[1]*For every 1 gram of the red clay particles, etc., that are attracted to the magnet, there are 2.5 grams that are not attracted.*
[2]*Must shape into a round tower to measure susceptibility.*

---

must be of cosmic origin. The earth, of course, is the center of a focused magnetic field called by geophysicists the *magnetosphere*. The sun-pumped solar wind *blows the magnetosphere* across the rotating earth from the day-to-night side and the force lines of the magnetosphere focus at the poles. The interaction of this mighty magnetic wind on temperature, winds, radio emissions, gases, atmospheric ions, etc., is too complex to discuss here. Suffice it to say that the sun is most certainly the transmitter of the magnetic energy to which my stone towers resonate. The ancients of course knew this, and that is precisely why so many of the ancient religions included stone structures in their worship of the sun god—indeed the sun god was common to many ancient religions.

Everyone knows that a magnet is a dipole (two poles) structure—N and S or + and –. A dipole, of course, is neutral—that is, one pole cancels the opposite so that it has an electron charge of zero. In 1931, P.A.M. Dirac, an English physicist, predicted, using quantum theory, that there existed such a thing as magnetic monopoles—that is, magnetic charges that do not cancel each other and thus emit magnetic waveforms. Recently (1981) Dr. Freeman Cope of the Naval Air Development Center demonstrated—utilizing modern sensitive equipment—currents of magnetic charge in flowing water. Other than the implications of such an elegant proof to dowsing, it should be obvious to even the most skeptical that since there is magnetic monopole wave function, then my experimental stone antennae and soil work is not only an experimentally demonstrated fact, but likewise rests solidly on the same particle-wave duality that was long ago established for electrical charges. I leave to my reader the implications of designing stone or clay paramagnetic wave resonant intensifiers for the stimulation of better crop growth on the farms of America. That, after all, is what I get paid for—to help farmers grow better crops cheaper. That is a truism, rather it involves insects—I am an entomologist—or clay and rocks.

7
THE DETECTION OF MAGNETIC MONOPOLES AND TACHYONS — A PICTURE OF GOD

Dr. Freeman Cope, now deceased, formerly of the Biochemistry Laboratory, Naval Air Development Center at Warminster, Pennsylvania, has postulated that man dwells in a gas of tachyon, magnetoelectric dipoles. Before his death, this brilliant physicist-M.D. and I spent many hours on the telephone discussing just such a possibility. His sudden death last fall was a real loss to the scientific world. His work in several areas, particularly the study of magnetism, deserved Nobel recognition.

Despite Dr. Blas Cabrera's (Stanford University) claim for the detection of a single magnetic monopole event, Dr. Cope (*Physiological Chemistry and Physics,* volume 12 (1), 1980) was the first to detect magnetic monopoles. He detected monopole currents in flowing water using what is known as a Josephson Junction solid state detector.

What is a magnetic monopole? Quite simply it is one or the other end of a magnet, north or south, + or −, all by itself. It should be quite obvious to researchers that if one can split an electronic field into two polarities, one should be able to split a magnetic field into its two polarities.

After all, the electric field has been understood to split into negative electrons and positive positrons since the early days of the cathode ray tube. The screen of your TV set is a perfect electron detector—it is the electron beam that makes it glow.

When one cuts a bar magnet in half, one merely gets two smaller bar magnets. The two halves are still + and −, or north and south. There does not seem to be any way to separate them despite the irrefutable fact that—just as in the case of the electric field—quantum mechanics predict separate north and south magnetic parts. The English physicist P.A.M. Dirac, utilizing quantum mechanics, predicted magnetic monopoles as early as 1931 (*Proceedings Royal Society of London*, A133). He never detected them, as far as I can determine, and gave up trying.

Most researchers agree that in order to detect magnetic monopoles, one must use a solid state detector that is superconducting. By superconducting one means that the circuit is cooled down to a temperature where the circuit has no electrical resistance to the flow of current, therefore the electrons flow completely unimpeded. Theoretically, if one could keep the circuit cooled forever, the current would flow forever.

What did Dr. Cope mean when he wrote his paper, *Man in a gas of tachyon magnetoelectric dipoles?* In simple terms he meant that we are bathed in a field of north and south magnetic dipoles (two poles), but that under certain conditions, each dipole is capable of dissociation into a pair of separate monopoles (+ or −), and that some of these monopoles are tachyons. *A tachyon is a particle with a speed faster than the speed of light.* Tachyons have been postulated by many physicists with enough imagination not to restrict their thinking to the idea that all wave or quanta phenomena are limited to the velocity of light (186,000 miles per second). I shall quote Dr. Cope directly as to the many little-understood phenomenon that might be explained by the monopole-tachyon theory. He stated in his paper:

"The biocosmic phenomena at issue in this paper include the following: (A) Diurnal cycles in antigen-antibody reactions at magnetic electrodes, apparently due to magnetic radiation of an unknown type from the sun; (B) Rays from the sun of unknown type which can be detected by sick (and by some sensitive normal) humans, absorbed by metal plates, and conducted by metal wires; (C) Rays from the sun of unknown type which affect human health and can be trapped in bilayer metal boxes; (D) Colored clouds (auras) around magnets and man which can be seen by sick people (and by a few sensitive normals) and can be made visible to normals by chemical sensitizers; (E) Grid lines parallel to lines of latitude and longitude, which are observable by certain persons (dowsers) but not by conventional magnetic or electric detectors."

It is a hang-up of modern science that many of these phenomenon listed by Dr. Cope, although proven time and time again to be true physical phenomenon, are regarded by many scientists to be delusions of the occult or mystic mind.

In 1979, I set out to see if I could detect magnetic monopoles. Why should an entomologist—with no budget at all since I did all the work on my own time with my own electrometer—believe he could detect monopoles when some of the best physicists in the country have failed using hundreds of thousands of dollars worth of equipment?

Quite simply, this was because I have not developed the contempt for God's organic nature that seems to envelope modern-day, high-energy inorganic science. I detected monopoles with a four-dollar *Ficus benjamina* (weeping fig) houseplant—simply an organic antenna hooked to my Keithley electrometer. I also, as Dr. Cope predicted, detected tachyons.

My reasoning goes back to work that I and my friend Dr. Ernst Okress did in 1965. I had published a couple of papers on insect spines (sensilla) as dielectric antennas for resonating to coherent infrared radiation emitted from plant scents and insect sex scent molecules. Based on my antenna/coherent-IR papers, Dr. Okress wrote a paper theorizing that such elegant systems are natural (room temperature) organic superconductors. Several other imaginative physicists (*Scientific American, Superconductivity at Room Temperature,* 212, 21, 1965) have put forth the same theory.

Since I agreed with the superconducting theory, the next step was quite logical. Why not use a room-temperature superconducting plant as an antenna-detector for magnetic monopoles and tachyons? Accordingly, I began continuous recordings in 1979 with the *Ficus benjamina*. Why that plant? Because it is a sturdy plant and keeps its leaves all year round. Also the *Ficus* plant is quite sensitive to both light and to being moved, so it obviously responds quite readily to its environment. Unlike my previous plant antenna work where I connected to the stem (see my book

Tuning in to Nature), I connected the electrometer between a leaf and the ground.

According to physicists and their complex mathematical analyses, the detection of a burst of magnetic monopole energy should produce a sudden increase in the current flow. However, we are talking in terms of 10^{-11} (only .00000000001) amps—a very minute signal!

Since my Keithley electrometer reads down to 10^{-12} amps (one trillionth of an amp), I was certain I could detect any sudden change in current. The mathematical description of such a sudden change in continuity is called a delta function. A delta function will describe the beginning of a square wave like this:

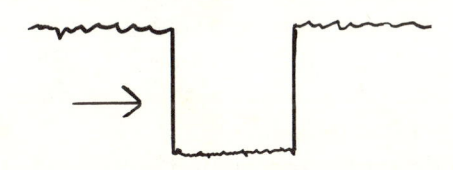

This is the same signal that Dr. Cabrera at Stanford received during his single event. At this point it might be well to distinguish Dr. Cabrera's monopole from Dr. Cope's monopole—they are two different sizes of the same monopole theory. When Dirac postulated magnetic monopoles in 1931 based on quantum theory, he was speaking of the existence of the + or − particles of the smallest charge. An analogy would be electrons for the electric field. Of recent times, physicists have put forth the grand unified theory and are looking for giant particles (10^{-8} grams). If there are such giant quantum particles, then it is likely that they are very rare and Cabrera's expensive, helium-cooled, superconduction detector only "sees" a small part of such an extremely huge magnetic quantum particle (wave). The event of Dr. Blas Cabrera at Stanford University may or may not indicate such a rare giant monopole, but my work backs up Dr. Cope's contention that monopoles and monopole tachyons are not at all as rare as we would be led to believe, and are what Dr. Dirac was describing originally. After all, with my superconducting *Ficus* plant detector, I recorded 165 lightweight monopole events over a two-year period and also 28 tachyon events (see table, page 83). How do I know I detected particles faster than the speed of light?

First, the tachyon events were associated with the cosmic generation of monopole events as predicted by Dr. Freeman Cope.

Second, they demonstrate what mathematicians call a recursive network structure and processes. Time lines on the best recording, of which there were seven (see table, page 83), either cross over or *kiss* one another, forming *orthorhombic* cells as predicted by H.C. Corben of the University of Toronto.

The mathematics of plotting recursive networks on graph paper is quite complex, but in simple terms, a recursive structure is one that repeats itself in smaller and smaller terms to infinity. We have all seen a painting of, say, a house which has a painting of the same house on a wall, having the same picture on its wall, etc. We can imagine that going on forever. That is exactly what the figure on page 81 (my best recording of June 24, 1980) demonstrates and that is what recursion is. From right to left or left to right, each *orthorhombic* cell repeats itself until it trails off to nothingness. Another characteristic of graph-plotted recursion is that they form mirror images of one another. You will note that the pattern repeats itself on the right and left side.

On both sides the lines appear to cross over one another. But to really understand why the recording is so astonishing, one must understand the mechanics of the recorder that produced it between 10:00 p.m. and 2:00 a.m. the night of June 24, 1980.

The instrument is a small Rustrack recorder hooked to the output of my Keithley electrometer. The *Ficus* plant antenna is hooked to the input of the electrometer. The recorder has a drive motor that rolls the paper from right to left. It is a single-direction motor and cannot reverse its roll, that is, it cannot start running backwards. Each second a sharp-edged bar strikes a galvanometer needle (the same as any volt or amp meter) and presses the needle against a round bar fixed in place under the paper which rolls across the bar.

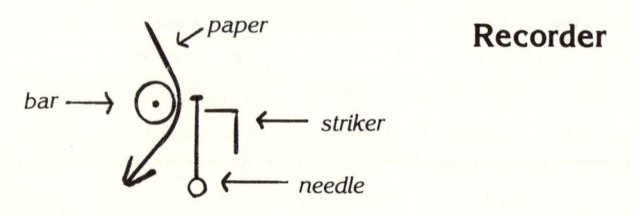

Recorder

Neither the bar nor the striker can move, but the needle moves back and forth between the two, depending on how many microamps of current flow in the galvanometer coil. When the striker hits the bar, it prints a dot on the pressure-sensitive paper depending on where the needle is—which, of course, depends on current flow.

Since each dot represents one second in time, and the chart paper rolls up four 15-minute increments per hour, it would be mechanically impossible for the needle to print dots backwards in time whether or not the time lines crossed as illustrated:

$$\times$$

or *kissed* like this:

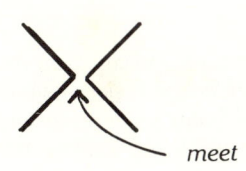

$$\times \leftarrow meet$$

The recorder roll only travels forward!

The recorder could not, therefore, form orthorhombic cells unless... unless what? Unless the *Ficus* plant is not only a superconducting detector, but is also a very sophisticated computer capable of processing and programming the incoming signals (registered by current flow as on any antenna) in such a way that time lines appear to be crossing or kissing one another.

The mathematical analysis that allows a computer-recorder to print out a continuous wave pattern like this:

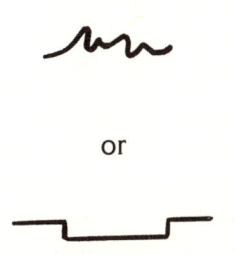

$$\sim\!\sim$$

or

$$\sqcap\!\sqcup$$

Magnetic Monopoles and Tachyons

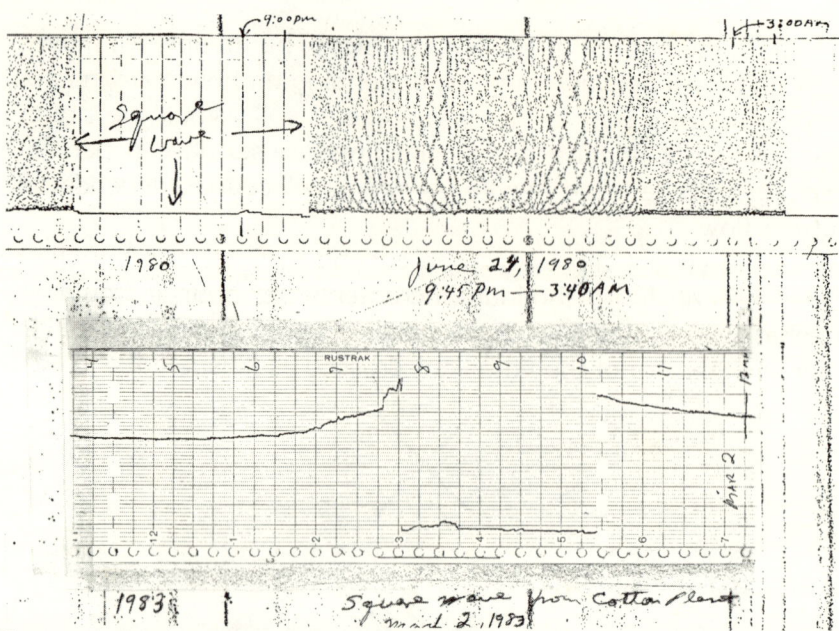

Top, a monopole square wave indicated by the space with the line at bottom followed by the orthorhombic, recursive network of a tachyon event. Below, a recording of a monopole square-wave event recorded from a cotton plant. The young cotton plant, in the bud stage, is an excellent magnetic monopole detector.

or some such single-line wave, is called a Fourier transform. For a signal to appear to reverse itself and go backwards in time, it would have to start and stop the recording needle so that the one second printout dots were placed on the paper in a very complex time-ordered sequence like this:

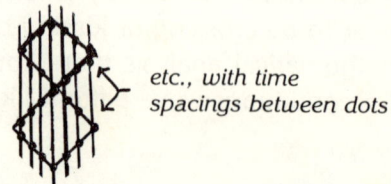

etc., with time spacings between dots

with time spacings between dots and in closely aligned, forward-moving, parallel time lines. This, of course, is impossible—or a miracle—unless there are two signals arriving at the plant detector, and one proceeds the other in time. (This is due to the mathematical nature of a Fourier Transform which is beyond the scope of this chapter.) The plant detector-computer then performs the normal complex Fourier transform. In other words, one signal must be

80   Ancient Mysteries, Modern Visions

a regular monopole signal (speed of light) and the other signal a monopole tachyon signal (faster than the speed of light) as predicted by Dr. Freeman Cope in 1980.

The only other explanation is that the plant has two different monopole detectors built into itself, *i.e.*, stem and leaf, an unlikely configuration for consistently efficient nature. One of the detectors would have to possess a time-lag circuit. This is highly unlikely because the electrometer was only connected from one leaf to ground and not from stem and leaf to ground. Furthermore, although there are two-detector, time-delay systems in nature, *i.e.*, an owl's ears—they are never connected to one lead (nerve), to believe a single, smooth leaf to be a double detector, one would have to postulate a totally unknown type of configuration and detector phenomenon. This would be presumptuous on the part of a biologist when physicists have not only predicted tachyons, but

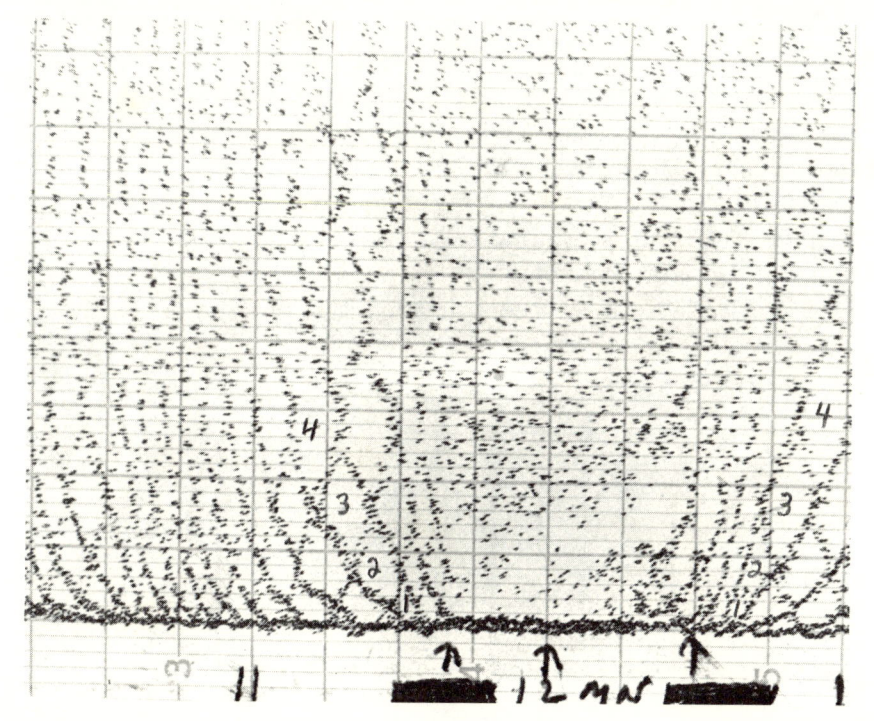

Orthorhomboidal, recursive pattern of the printout. As you can see, the rhomboidal shapes (1 to 4) get smaller and smaller and disappear into the baseline. Recursiveness, rhomboidal shape and square-wave association are three very exacting criteria for the deduction of tachyons—all three occur in the Ficus leaf reading.

outlined the exact mathematical configurations that their detection would encompass. I obtained that exact and very complex printout from my friend the *Ficus* plant.

The inescapable conclusion is that:

1. Tachyons do exist.
2. They occur, as predicted by Dr. Cope, in conjunction with weak magnetic monopoles.
3. They are easily detected by living plants.
4. Therefore, living plants (of interest to farmers) are indeed superconductors.
5. An entomologist was the first scientist to detect tachyons.

Physicists have always maintained that if something mathematically ought to exist, then it probably does. It then becomes a matter of learning how to detect and manipulate such phenomenon.

Of what use to modern man, and especially to agriculture, is the knowledge that monopoles and tachyons actually exist? One thing that immediately comes to mind is that since a leaf is an antenna superconductor detector and computer, then the collection of monopole and tachyon particles by the leaf is probably tied to the miracle of photosynthesis. For my agricultural friends, I need to go no further than the word photosynthesis to emphasize the importance of tachyons to mankind.

As for myself, the very fact that my love of science has never

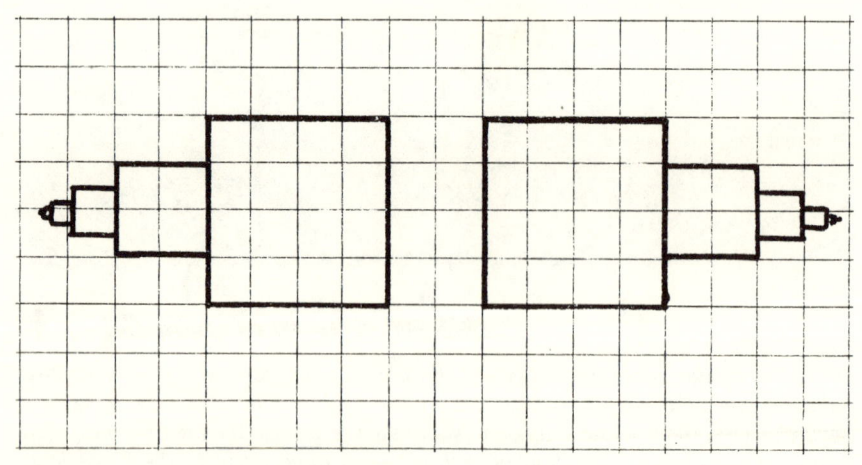

Example of a square recursive plot on graph paper.

TABLE 1
TWENTY-EIGHT TACHYON EVENTS (JUNE 3 TO 12 NO EVENTS), 1980.

No.	Date	Time (hours)	Comment
1	June 13	0815 to 1000	
2	June 14	0 events	
3	June 15	0645 to 0715	
4	June 16	0 events	
5	June 17	0 events	
6	June 18	0 events	
7	June 19	0730 to 1000	
8*	June 20	0200 to 0615	Good
9*	June 21	1015 to 1230	Good (summer solstice)
10*	June 22	1000 to 1730	Very good
11*	June 23	1230 to 1745	Good
12*	June 24	1145 to 1430	Good
13*	June 24	1745 to 1900	Good
14*	June 24-25	2145 to 0340	Very good
15	June 25	0945 to 1600	
16	June 26	0745 to 1600	
17	June 26	1830 to 2100	
18	June 27	0230 to 0915	
19	June 27	1430 to 1545	
20	June 28	1030 to 1230	
21	June 28	1920 to 1400	
22	June 29	0145 to 0745	
23	June 29	1250 to 1500	
24	June 29	1910 to 2045	
25	June 30	1310 to 1350	
26	June 30-1	1450 to 0400	All night long
27	July 1	1245 to 1322	
28	July 2	0010 to 0145	
29	July 2	1215 to 1345	
30	July 2-3	2330 to 0130	
31	July 3	0845 to 1015	
32	July 4	0930 to 1045	

Best recursive graphs occurred around the summer solstice—so tachyons appear to be associated with the solstice!

blinded me to God's own handiwork and has thus permitted me to be the discoverer of tachyons, is fulfillment beyond all expectations. This is so because, as mathematician-physicist Douglas R. Hofstadter has written in his book *Godel, Escker, Bach, an Eternal Golden Braid* (Vintage Books, New York), concerning a recursive plot, called a G plot:

"You might well wonder whether such an intricate structure would ever show up in an experiment. Frankly, I would be the most suprised person in the world if a G plot came out in any experiment. The physicality of a G plot lies in the fact that it points the way to the proper mathematical treatment of less idealized problems of this sort. In other words, G plot is purely a contribution to theoretical physics, not a hint to experimentalists as to what to expect to see! An agnostic friend of mine was so struck by G plot's infinitely many infinities, that he called it "a picture of God," which I do not think blasphemous at all."

As Dr. Hofstadter points out, no blasphemy was intended, only the recognition that God's handiwork is elegant beyond the wildest dream of mankind. Truth is stranger than fiction!

8
SAND FROM EAST TO WEST

"These labors consisted in tilling and planting and watering the fields, and in bringing sand from east to west and doing whatsoever had to be done in connection with agriculture in the other world."
—Osiris and the Egyptian Resurrection by E.A. Wallis Budge

Anyone who has ever visited a museum containing Egyptian artifacts must have noticed the mummy-like wooden figures so often present in exhibits. The ornate, beautifully carved, gold-coated figures are called *Shabti* or *Shauabti*. Sometimes they are made of stone or alabaster. I have one made of hard basalt rock. It is ancient and poorly carved, but then it is hard to understand how anyone carves basalt at all. I also have a more gaudy one of painted clay-covered wood.

If I were an ancient Egyptian of the 6th to 18th dynasty, these little figures would be buried with me and would serve as my agricultural hands in the next world. The Egyptian concept of paradise—for like we Christians they believed in immortality—was personal and very, very human. The Egyptians were realists.

The quote at the beginning of this chapter describes the duties of these laborers as inscribed in hieroglyphics on the base of Shabti taken from the tombs of wealthy Egyptians. Presumably these ancient landowners did not like tilling and hoeing in the hot Nile sun any better than we moderns might. Therefore, they provided for an easy life in the other world.

Most archaeologists are of the opinion—and the evidence is strong—that the earliest dynasty rulers sacrificed living humans in

86 *Ancient Mysteries, Modern Visions*

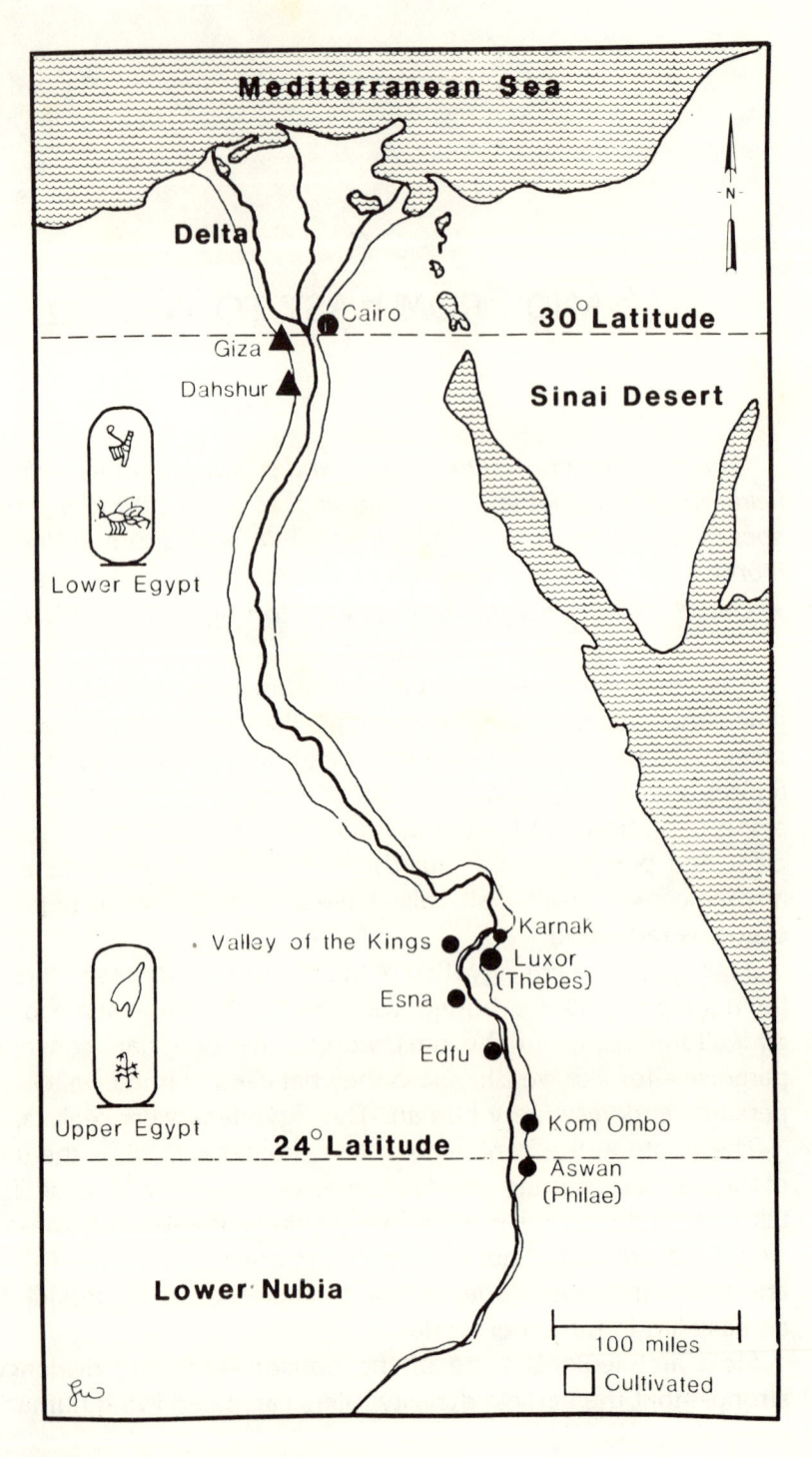

order to provide labor in the next world. As the Nile civilizations progressed, the little Shabti dolls were substituted for living persons—a far more elegant concept in my opinion. Instead of being shaped like mummies they began to be carved in lifelike style, usually holding the tools of their agricultural trade.

We can, of course, understand why a landowner passing to the other world would want laborers to help with the tilling, planting and watering, but why bring sand from east to west? For what purpose would they bother to ferry sand from the east bank to the west bank of the Nile? Nowhere in the archaeological literature is this strange behavior explained. In fact it is not even commented on as far as I can determine from a literature search.

My two Shabti are of the mummy form with no hands or arms showing. Many Shabti from the mid-dynasties are shown with each hand holding a hoe in hieroglyphics and a basket thrown over one shoulder. The basket is, without a doubt, for carrying the dirt to a barge in order to ferry the dirt to the west bank.

Throughout Egyptian history the west bank was called the bank of the dead. All of the tombs and pyramids are on the west bank. The east bank was, and still is, called the bank of the living. It has been presumed by those who study ancient Egypt that the west bank is called the bank of the dead because the dead were buried there. That may be a partial explanation based on what we know about the later dynasties, but it is, in my opinion, a highly unsatisfactory explanation as to why the west bank was originally called the bank of the dead.

The Egyptians, who as I have pointed out in previous chapters were not the starry eyed superstitious mystics we make them out to be, must have had some solid basis for considering the west bank dead—at least that bank across the river from Luxor.

It was on a trip down the Nile on a little riverboat called the Karnak that it first came to me—insight I believe it is called—why the Egyptians labored so hard carting sand about. The reason was that the land on the west bank across from Luxor was dead. I am quite sure that my farming friends will understand exactly what I mean by dead land, or dead soil, to put it more accurately.

The next question, of course, is why is the land on the east bank living? Dead land would mean dead agriculture and consequently dead people—*famine*! Living land must be living for some reason that is not given in any modern book on soil, nor is it obvious from what we can read in Egyptian hieroglyphics. However, as is usual when one spends years at a scientific endeavor, things will finally fall into place. They began to fall in place for me when, on a beautiful morning in May 1980, the Karnak pulled away from the dock at Luxor at 3 a.m. with my wife, myself and 30 Danish Egyptologists aboard.

I will never forget awakening that spring morning as long as I live. The diesel engine of the Karnak rumbled alive about 3 a.m. and by four we were sliding silently along the river bank past Luxor. In the dim light of the false dawn, I could see the ghostly pillars of Luxor temple against the red haze of the desert sky. It was a still, almost hauntingly silent morning with no wind to ripple my huge Bartholomeu World Travel Map of Egypt.

Most of my original map is dark brown or sandy in color, except for the long green line that borders the painted blue of the Nile. (Unfortunately, this map has to be reproduced in black and white in this text.) At the top, the green fans out into the funnel-shaped delta. Cairo is located at the apex of the triangle where the fan-shaped agricultural delta land narrows to a ten-mile strip on either side of the river. For about 300 miles the river makes a gentle curve toward the southeast. It then suddenly turns due east and quickly loops back to the west. That is the big crook in the map's blue and green neck. It is at the south end of this huge 80-mile-long crook that Luxor is located.

In ancient days Luxor was called Thebes and was the religious and political capital of the middle dynasties. Their glories are the Luxor and Karnak temples and, of course, across the river on the dead bank, the Valley of the Kings with the fresco-decorated tombs of the pharaohs cut into the huge two-hundred-foot-high cliffs. King Tutankhamen was lifted from one such rocky tomb.

As the day progressed I kept track of the villages and landmarks on my map. With the exception of considerable land around the town of Isna, 50 miles south of Luxor, much of the farmland lies on the east bank. The main road is also on the east bank. Below Edfu, where the temple of Horus the sun god is located, all of the good farmland lies along the east bank. At Edfu it is green on both sides of the river, but farther southward, past Kom Ombo where the temple of the crocodile god is located, the good land is again

88   Ancient Mysteries, Modern Visions

spread along the east bank all the way to Aswan.

The first rapids of the Nile are at Aswan and boats cannot travel south of that city. The huge Aswan Dam blocks the river and forms the 200-mile-long Nasser Lake behind its massive Russian-built concrete form.

I call the Aswan Dam "the Strangler." Why? For the very simple reason that it cuts off all of the river Nile from the soil "life giving" forces that for 6,000 years of Egyptian civilization flowed from in-

The trinity of ancient Egypt. Osiris, the father god, wears the crown of upper Egypt. He was dismembered by his evil brother Seth and later reassembled by Isis, the mother-sister god, represented here with the two horns of the sacred cow and holding the sign of life (oval and cross) pronounced ankh, meaning life in the Egyptian language. Horus, the powerful falcon and sun god, wears the crown of unified upper and lower Egypt. Isis and Horus resurrected the dismembered Osiris by putting his body back together. The cycle of life is resurrected when the dismembered (dead) organic matter of the White Nile joins the powerful paramagnetic silt of the Blue Nile to begin again the cycle of agriculture at the diamagnetic water of the Nile proper.

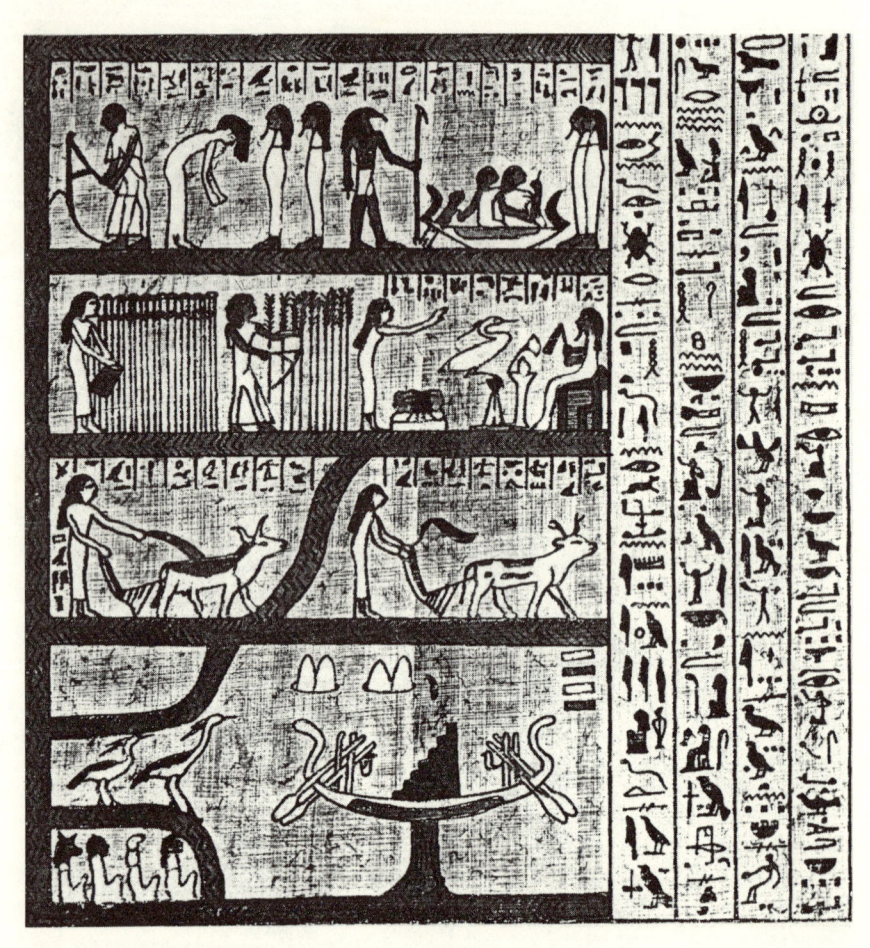

The importance of agriculture to the Egyptian civilization is seen in this beautiful papyrus drawing of Lady Amhai paying homage to her dead parents. All the scenes are agricultural in nature. Note that the first drawing is of a man with a hoe, symbol for the verb "to love."

ner Africa. These are the earth forces, now lost in memory, that the pragmatic ancient civilization of Egypt honored through its religion, and also protected with its politics and its armies.

The Egyptians eventually evolved an elegant and practical religion based on a primary trinity of gods—Horus, Isis and Osiris. I believe that the three gods for centuries represented the stone- and water-generated lifeblood which contains these forces. On my river journey I had slowly come to understand the subtle meaning behind the concept of the trinity. There was Horus the male sun god, the source of the stone paramagnetic energy; Isis, the gentle mother and life sustaining force of diamagnetic water; and the father god Osiris who was cut apart, according to Egyptian legend, and his organs and body scattered about the Egyptian countryside by his evil brother Seth (a sort of devil figure). The gentle mother Isis, and powerful son Horus gathered up the several parts and reassembled them.

The fascinating history of the development of the powerful Egyptian trinity—son Horus, father Osiris, and Isis the sister and wife of Osiris—is very complex and evolved over the centuries from many minor gods of great diversity. Suffice it to say that Osiris became the best loved god of the middle dynasties.

The elegant doctrine of Osiris held that it was Osiris, the father god, who introduced both art and agriculture into Egypt. As J.C. Manship White has written in his marvelous book *Ancient Egypt*, "He was also a nature god, associated closely with the river and the rich soil of the Valley. He was the principal of fertility, in opposition to the principal of sterility personified by Seth."

How did those three godlike forces—paramagnetic soil, diamagnetic water, and organic matter—become so much a part of an elegant religion that lasted for centuries?

The art, history and literature of ancient Egypt indicates that here was a highly realistic agricultural civilization with a well-defined belief in trinity, incarnation and sin, heaven, hell and salvation, and also a belief in the death and resurrection of a god, Osiris. Although the details may differ, the basis of the Egyptian beliefs were little different than my own Christian beliefs, which believes in, guess what, a trinity, incarnation, sin, heaven, hell and the salvation and resurrection of a God man.

In the book of the dead, as translated by the brilliant English Egyptologist E.A. Wallis Budge, we find in the segment on Osiris, as judge of the dead, the Egyptian declaration of innocence before entering paradise. There are forty statements in all. For example, they read:

Hail, Am. Khaib tu, coming forth from Qerrt, I have not committed theft.

Hail, Neha. Hau, coming forth from Re-stou, I have not killed men.

Hail, Uamemi. Coming forth from the House of the Black, I have not laid with another man's wife.

Sound familiar? Moses, an honest prophet, would agree. It is only high-energy oriented modern man who looks down his condescending nose at the ancient Egyptian civilization and considers it a sort of occult assembly of cruel slave traders.

If the Egyptians were so heartless, why did Christ flee into Egypt instead of, say, Syria? I would assume that it was because although Jesus, Joseph and Mary were Hebrew, they were welcomed in Egypt. The traditional Synagogue of the Holy Family still exists in Cairo. Even if it is not the original temple, although it may well be, at least the tradition has lasted over 2,000 years. The ancient Egyptians only conducted about one war every century. If only Europe and the United States had a score like that.

Most Christian tourists consider a visit to the Holy Land a must, but if they really wanted to understand their religion, they would do well to include Egypt as a holy land and visit that country. Fortunately today it is possible.

Where did the three god forces of ancient Egypt come from? From the same place, according to knowledgeable Egyptologists, that Egyptians themselves came from—inner Africa. The paramagnetic god Horus flows down the Blue Nile from the volcanic highlands of Abyssinia. The organic god Osiris flows from the green, inner-jungle lands of central Africa. Osiris is the dead greenery and must be found by the powerful sun god Horus (paramagnetic soil) and gentle mother god Isis (diamagnetic water) to be resurrected again. The resurrection occurs in the Sudan where the Blue and White Nile join. The elegant trinity of forces join together for the resurrection of the cycle of life in the Nile Valley. The Egyptian natural trinity can be easily seen in their ancient symbol for life—a sort of cross and circle.

How have we moderns managed to almost destroy that Egyptian trinity of agricultural forces?

Although the Nile Valley is quickly losing its fertility due to the excessive use of salt fertilizers and the dumping of tons of pesticides on the cotton crop—three times as much as on the lands of the American south—it is less the quick-fix, high-energy modern farming techniques that have ruined Egyptian agriculture than the building of the *Strangler Dam*.

The damage done to Egyptian agriculture by the placement of

Hoe   **Ankh**

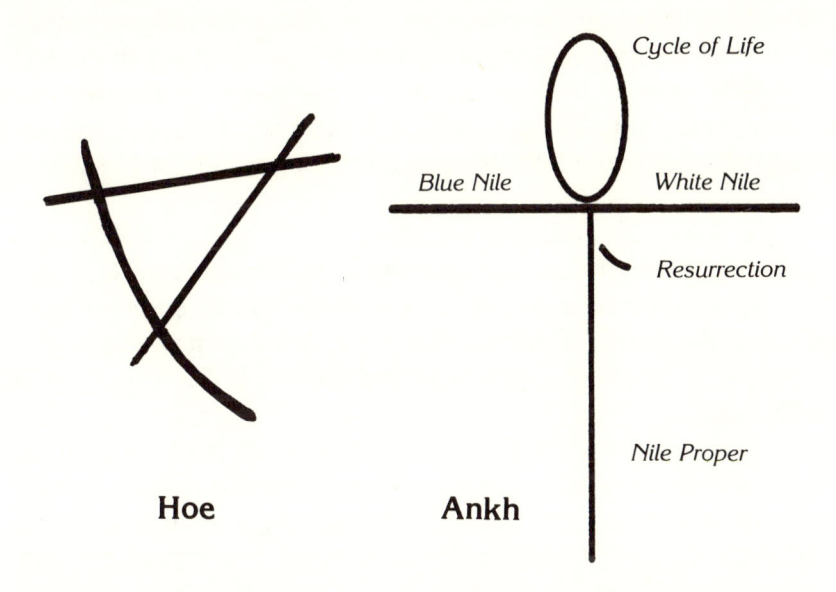

- Cycle of Life
- Blue Nile
- White Nile
- Resurrection
- Nile Proper

the Aswan Dam across the Nile River could have easily been predicted, and a few agricultural ecologists did indeed predict it—but that is entirely another story. In simplified terms, the destruction of Egyptian soil was begun when the huge High Dam of Aswan blocked the flow of the river and trapped the rich alluvium behind its massive bulk. The rich fertile mud no longer reaches the Egyptian Nile. The flow of Nile blood has suffered cardiac arrest.

The lands of the Nile Valley are unique because they are automatically, so to speak, fertilized by the June inundation each year. This natural fertilization is superior to that of almost any other valley in the world. It is superior because of the two Niles feeding the Nile Valley south of Khartoum, Sudan. The Blue Nile has its source in Lake Tana in three rivers—the Atbara, Dinder, and Rahad—that flow from the mainly volcanic mountain highlands of Abyssinia. The Abyssinia Mountains are almost entirely of volcanic origin and as emphasized earlier, volcanic rock and ash are highly paramagnetic. Volcanic rock erodes into volcanic soil.

Due to the mountains, the southwest monsoon creates enormous flooding and also considerable soil erosion along the feeder mountain rivers of the Blue Nile. In other words the fertility of the Egyptian Nile north of Khartoum depends on the erosion of its Abyssinian highlands. Egypt's gain is Abyssinian's loss!

Sand from East to West   93

The White Nile, on the other hand, is a slow-flooding river that has its origin in the lush jungle forests of Lake Victoria in central Africa. It brings with it a rich mulch of organic matter. By the time the Abyssinian floods have passed Khartoum, they are slowed considerably and joined to the mulch-carrying waters of the White Nile.

As the June floodwater slowly rises along the Nile Valley, the Egyptian farmers with the forewarning of the temple priests, moved to higher ground and awaited the three month summer subsidence of the floods. During this summer break in growing crops, the people engaged in other activities such as building pyramids and temples.

The Abyssinian highlands are eroded by winds the year around, and the monsoon cloudbursts are so unpredictable that the priests no doubt had a very difficult time forecasting the advent of the flood each year. Even with their great knowledge of astronomy and the seasons, they could not have been accurate too often—or were they? Edward Hyams, in his masterpiece, *Soil and Civilization*, quotes the famous German Egyptologist Emil Ludwig from his work, *The Nile*.

"So many people and generations of Egypt have studied this vital question through and through, and yet the height of the flood resulting from the rain [in Abyssinia] has never once been forecast for the following years."

Of course, Ludwig could be mistaken. It well may be that we moderns, with our poor understanding of weak natural forces, cannot predict the floods. It is quite possible that the ancient priests did predict them, and were thus able to promote political stability for century after century.

Unless we moderns put more research effort into the study of low-energy organic physical forces, such as infrared radiation, paramagnetism and diamagnetism, we may never understand the so-called magic of these ancient peoples. Be that as it may, the Egyptians had a very good way to promote political stability, and that was with a single religious belief that unified entire peoples. Most certainly the ability of priests and pharaohs to prove incarnation by levitating a body, alive or mummified, would strengthen a people's faith. One does not need to read a complex theological tome to understand that possibility. The pyramids were temples for just such a levitation ceremony, as I have noted in previous chapters.

As the Karnak stopped at different places along the Nile, I collected sand from the desert in little plastic bags. When I arrived back at my hotel room, I would fill one-quarter-inch-diameter glass tubes with the sand. A glass tube is neutral, neither dia- nor paramagnetic. The sand-filled glass tube was then hung from a thread in the manner of my round tower sensors. In every single case where I had collected sand along the Nile and there was no agriculture, the sand-filled glass tube was either neutral or repelled (diamagnetic) when I brought my 2,000-gauss cobalt magnet close to the tube. Wherever there was agriculture, the sand was highly paramagnetic, because it was composed of eroded paramagnetic volcanic grains of earth.

Today, along the northern reaches of the Nile between Luxor and Cairo, there is more good agricultural land on the west bank than the east bank, but it is impossible to tell how often the river has changed course from one side of the Nile Valley to the other or how much good, paramagnetic sand was carried to the west bank. Certainly, however, in the region of Luxor, Edfu, Kom Ombo and Aswan, the agricultural land remains even today on the living east bank of the river.

It is during the temple building era, in and following the 18th dynasty, that the hieroglyphic writings called *The Book of the Dead* is dated. The height of Egyptian civilization is considered to have occurred during the middle dynasties, and it is here that we read that the Egyptians carried sand from the east bank to the west bank.

If, due to the nature of the seasonal floods in this region, most of the rich Abyssinian eroded stone and clay ended up on the east bank and one wished to grow crops on both sides of the river, what is more logical than to carry the paramagnetic soil and clay from east to west. In truth, thousands of years of carting soil to the west bank may as easily explain the present day good soil on both sides of the river.

Modern institutional scholars may scoff at such a thesis, but modern urban scholars have lost their *feel* for the soil. They are consequently apt to attribute some useless occult reasoning to everything written in the *Book of the Dead*.

I know that if I were an ancient Egyptian crossing the Nile River to my garden in the sky, I would want some help carting the paramagnetic soil of my Horus god across the river so that I could plant a garden paradise on the dead west bank. Without that good growth force, my west bank paradise would be but a hellish desert. I am certain that my little Shabti dolls would be more than happy to help in my heavenly garden project, and that is why I am firmly convinced that the highly pragmatic Egyptians had a very good reason to carry sand from east to west, whether in this world or the next.

9
Monopoles — To Love the Land

I believe that there is one sure way that I can prove to my reader that the ancient Egyptian loved the land and his soil above all else.

The Egyptian hieroglyphic symbol for a hoe is a copy of the actual tool that the farmer used to cultivate his soil. It takes the shape of a curved pulling stick with triangular handle attached to the top. It is seen as the very hieroglyphic in the papyrus tribute of Lady Amhai to her dead parents. The hieroglyphic for the noun *hoe* also has a second meaning as a verb. The word *to love* is the very *same* hieroglyphic! I rest my case!

The Egyptian hieroglyphic language has been well preserved because it was inscribed into hard, paramagnetic stone.

Egyptologists, especially the famed Sir E.A. Wallis Budge, have done a superb job of translating ancient hieroglyphics. Beginning with the brilliant work of Akerblad Young and Champollion le Jeune and the famous optical scientist Dr. Thomas Young's translation of the Rosetta stone, Egyptian hieroglyphics have become almost as easy as any modern language to translate. But their real meaning is an entirely different matter.

Most hieroglyphic symbols have a double meaning e.g. ⌒ equals mouth, but it is also the symbol for the letter *r*. Y__Y equals the breast and arms of man, but it also means the syllable *ka*. Thus Egyptian hieroglyphics can stand alone or can be put together into syllabic words. The word for hoe and to love ⚹ , also carries

Monopoles 97

the syllabic meaning *mer*. Besides the actual and syllabic meanings of each hieroglyphic, there is, in my opinion, a third, hidden meaning in the Egyptian language. It is an actual picture language of the agricultural knowledge of those ancient soil-loving peoples.

Years ago I discovered that Ireland is composed of two types of limestone. One type is highly paramagnetic and is found in the mountainous rim of Ireland. The other type is highly diamagnetic and is the subsoil limestone bedrock that underlies the entire central agricultural bowl of Ireland. Both look exactly the same to the most practiced eye. That diamagnetic bedrock is, of course, overlaid by a rich top covering of highly paramagnetic volcanic soil which eroded down into the bowl from the volcanic rocks of the rim mountains. Ireland, like the Nile Valley, was perfectly designed by God for efficient agriculture—that is before our high-energy insanity took over to ruin it as in the case of the Nile Valley.

Strangely enough, all of the round towers and tower houses that I measured are made out of the highly paramagnetic limestone, even though they are located in the central bowl miles away from the mountainous limestone. Conversely, all of the cotter houses or old peasant cottages that I measured are made of diamagnetic limestone regardless of where they are located. Logical? Fighting and ruling is a charged up, fast moving game, whereas farmers are more *easy going* and like to relax in their homes after a long day's work with their paramagnetic soil. Those Celtic peoples most certainly knew about the opposite forces in stone, and probably inherited that knowledge from the ancient Egyptians.

The Egyptians had two hieroglyphics for stone—both take the exact proportions of the building stones found in great pyramids. One hieroglyphic is open like this ⊏⊐ , the other has lines across it like this ⊞ , the same way my model paramagnetic model round tower has force lines across it (see photograph, appendix 2). Both symbols represent the same syllable for stone (aner). Now you may understand what I mean by a secret agricultural language. Even the Egyptian hieroglyphics tell us that there are *two kinds of stone*.

The Egyptian hieroglyphic for *prepared stone* is aner sept (two syllables). The hieroglyphic is:

98   Ancient Mysteries, Modern Visions

It contains a feather for levitation, waves, a mouth (source of paramagnetic breath), a stone (with paramagnetic force lines) and finally a pyramid (Septih, the dog star). The little circle is the sign for sand from which rocks are made, the bar ■ the sign for symmetry, and the three little lines ||| for plural (many building stones). Now you see why hieroglyphics are so easy to read directly.

The hieroglyphic for black granite is:

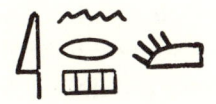

This word is similar to prepared granite except for the pyramid sign (Septih), for the dog star, is replaced by the symbol for a wing (many feathers), a much stronger levitating force than one feather ⟨. In other words, black granite is a paramagnetic battery for the force. Every Egyptian word for different types of paramagnetic stone *e.g. aner-en-bekhenu* (porphyry) of my Hueco Tanks, *aner-en-moat* (stone of truth), *aner-en-rut* (sandstone), *aner-hatch* (white limestone) etc., all have the symbol ⊞ with force field lines in it.

In words that are general, and not specific for a paramagnetic source, *e.g.* the word *weight* (all stones have weight), there are no force lines. The hieroglyphic for weight is:

a seed
a sieve   loop for scale

Symbols such as that for stone ▭ are called determinatives because they determine the main meaning of the word. In that way the Egyptians connected the syllabic meaning, *e.g.*, aner or stone to the object itself. Sieved seed is weighed.

Monopoles 99

Again, I rest my case for my thesis that the ancient Egyptians knew far more about low-energy physical systems in nature than we do today.

Knowing what we now know about para- and diamagnetism in stone, and magnetic monopoles and tachyons, can we now make a model of what these natural forces are all about—a magnetic model so to speak. Models are, of course, a legitimate part of science. The atom is merely a model of the form that we believe the smallest whole particles of nature takes. No one has ever seen an atom, only the results of what atoms do to a photographic plate. The same hold for our models of waves, X-rays, photons of light, etc. Just as you can't *prove* an electron or atom unless you *see* it, you can't prove God, or even a miracle, except by *seeing* Him or directly witnessing a miracle.

There is a whole group working on what is purported to be Jesus' burial shroud, called the *Shroud of Turin*. They have determined it to be 2,000 years old, and the image is inexplicable by every interpretation of the scientific tests they have conducted. In the same way, I have studied a painting thought by the Mexican culture to be a miraculous image. I used infrared film and demonstrated that there is no drawing or sizing under the pigment, nor protective varnish over the image of the Virgin of Guadalupe. The image of the beautiful virgin is "stuck" to a cactus cloth called agave cloth. It is a type woven by the Aztec Indians and sewed into a *tilma* (Indian cloak). The tilma should have rotted within 20 to 50 years. In other words, like the *Shroud of Turin,* the beautiful Virgin of Guadalupe, hanging in the Basilica of Guadalupe in Mexico City, is inexplicable! Did I *prove* it to be a miracle by scientific techniques? Of course not! What it does is strengthen our faith that miracles are possible.

Of course the fact that, according to all criteria set forth by reputable scientists, I have detected tachyons tells us in no uncertain terms that miracles are possible. Why? Because time, as Einstein said, really is relative and the speed of light (186,000 miles per second) is merely an illusion of our limited senses, and therefore is not really a true measure of what scientists call time.

Both the image of the Shroud of Turin and the image of the Virgin of Guadalupe are purported to have happened instantaneously. We may understand then, according to the rules of the

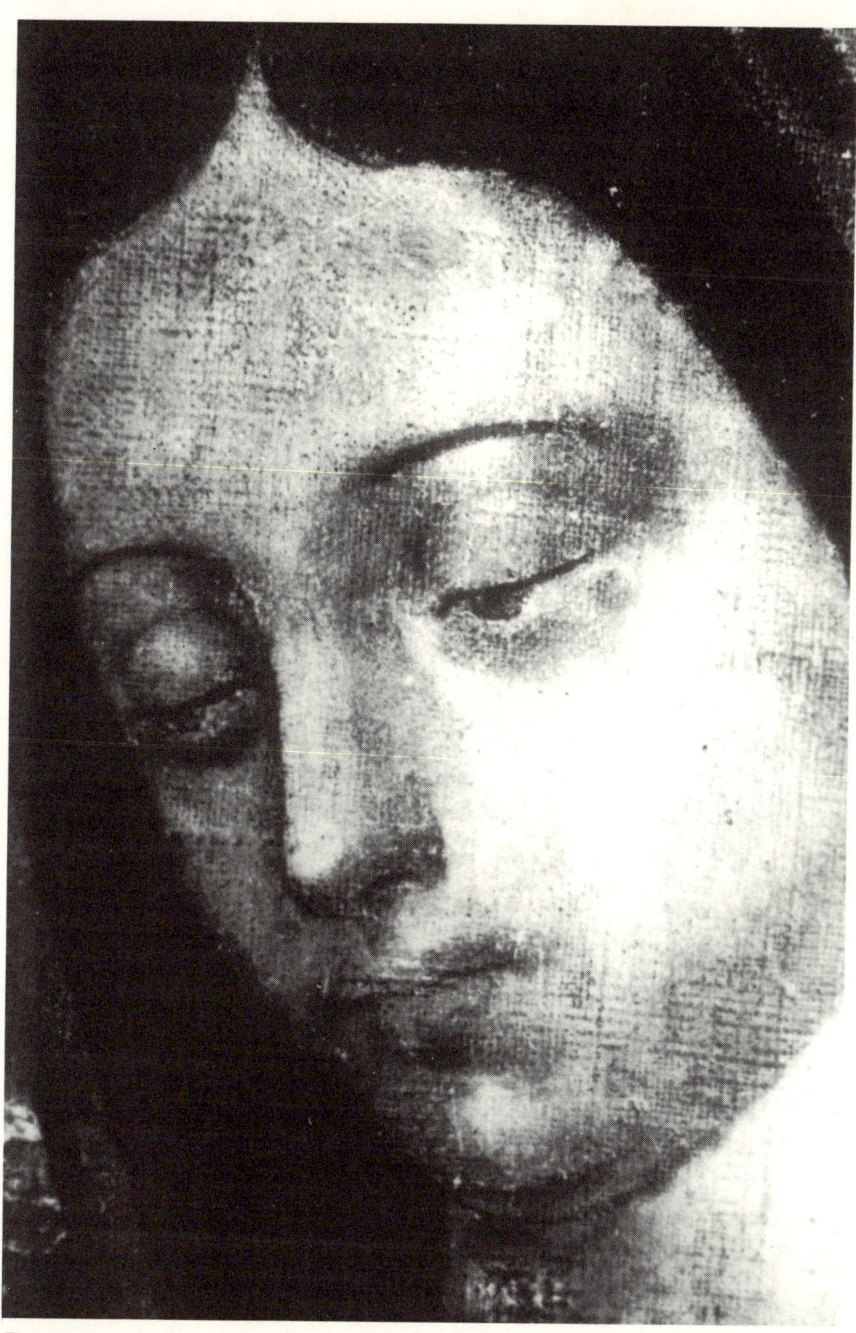

The beautiful image of the Virgin of Guadalupe hangs in a modern Basilica on the outskirts of Mexico City. It is considered miraculous by most Spanish-speaking peoples. My infrared study shows it to be inexplicable.

game called science, that since tachyons are a fact, then time can go forward or backward. In physics we describe such an event with a Feynman diagram. It looks like this:

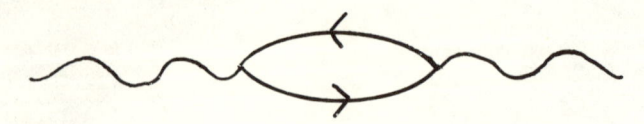

with arrows going both directions to show time *meeting* itself. If that sounds like a crazy scientific model, consider this: Professor Feynman won the Nobel prize for it! My tachyon work makes it seem that they won't have to take Dr. Feynman's Nobel prize back. Both he and Einstein are correct! If those two time arrows can meet one another, then time relationships, as our senses understand them, are relative. Therefore, it is possible for *any* event in nature to take place instantly—including the Virgin of Guadalupe and the Shroud of Turin. The fact that I picked up tachyons with my *Ficus* plant certainly strengthens my faith in both God and miracles!

As a biologist I believe that the theory of evolution is an elegant model for creation, but you will never hear me bad mouth creationists (as believers in instant creation are called), because if I did I would be denying my own tachyon research.

Perhaps God created life using *both* methods the same way I can produce a drawing in pen and ink or in watercolors. The entire controversy over evolution *vs.* creationism can only be categorized as stupid. It is indicative of the rigid religious orthodoxy of modern science. Both creationism and evolution are logical models of how we, with our limited senses, think life was formed.

So much for tachyons and miracles. How about soil and magnetic monopoles? Let us postulate a model for them. The model might turn out to be of utmost importance to my farming friends and to our own survival—*no food, no life and no America as we know it!*

Let us keep the model simple. Science always starts with simple models; it is only later that things get complicated. All great discoveries are based on simple models. Niel Bohr's original model of the atom was simplicity itself. So was the Wright brothers' model of an airplane. Look at a Boeing 747 today.

Let us begin with the magnetic model that A.R. Davis and W.C. Rawls, Jr. use in their book on magnetism (see bibliography).

The Davis-Rawls model considers that a magnet has a neutral equator (no force) in the middle, and that the force of the north pole spins one direction and the force of the south pole spins the opposite direction. They call north − and south +. I prefer it the other way around since I like to think of the North Pole as being positive (I live closer to the North Pole), but that is a matter of choice, and does not affect our model. In fact using symbols + and − for a magnetic force is confusing for they are already used for the electric force. To keep the model simple we will reserve + and − for *electric force* and N and S for *magnetic force,* also the end of the compass needle that points north is labeled south.

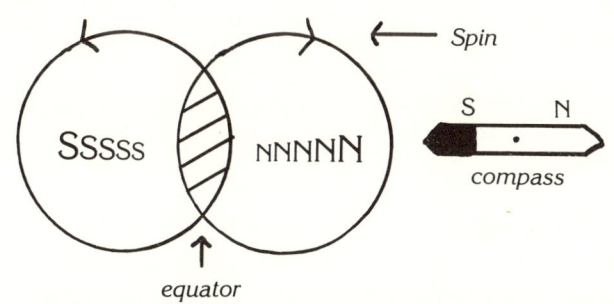

What we are talking about is not the physical magnet itself, but the magnetic field force. In quantum mechanics, however, you can mathematically consider forces as particles of energy (called photons for light). What we have then, in this Davis-Rawls model, are two opposite forces spinning opposite directions and bound together at the center. It makes sense that there is a neutral equator since opposite spins would run into each other at the center and cancel one another just like two cars running into each other.

This then is our *known* magnet at earth temperatures. But what if at extremely high temperatures, say the temperature of sun flairs, these magnetic forces are torn apart. My preliminary data indicates that more monopoles are detected by plants at high sunspot periods of activity. And since they occur at night and day, and since the earth is surrounded by a magnetic field, it makes sense that the action of sunspots is on our earth's magnetic field

Observed and one-year-ahead predicted smoothed sunspot numbers.

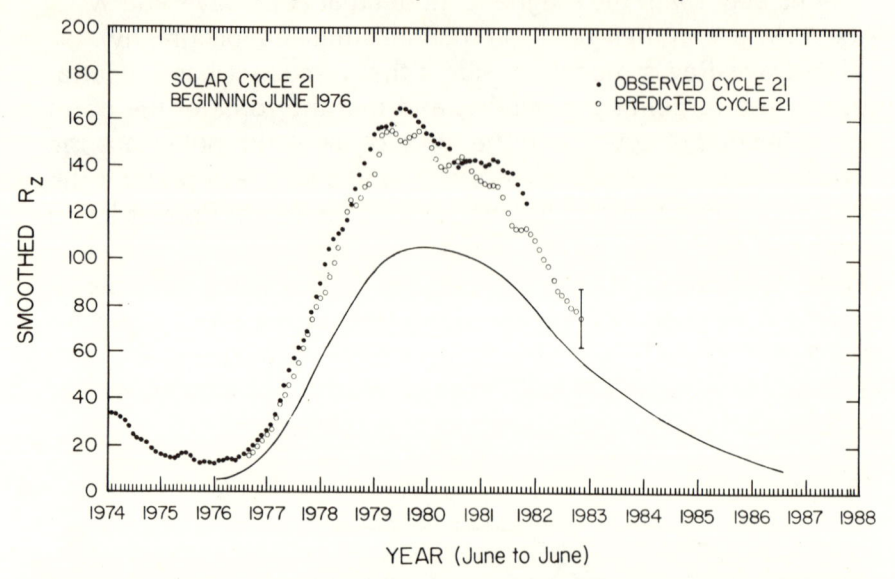

Sunspot cycle between 1974 and 1983 (dots) and projected cycle to 1987 (line). Magnetic monopoles and tachyons were at maximum at the peak of the curve in 1980 (see chapter 7).

and not with the earth itself. In other words, what I am measuring is a disturbance of the earth's magnetic field caused by monopoles emitted from extremely hot sun flairs. It is the *temperature* and *shape* of the flairs that tears the magnetic poles apart and sends them on their way in the same manner that heating a metal filament separates the electric field into opposite forces and sends out negative electrons or positive positrons. We can understand now that once these poles are torn apart they do not come back together, any more than electrons and positrons are reunited. They just wander around until they are absorbed by something. By what? Well, the south monopoles are absorbed by stone and soil, and the north monopoles by plants.

Let us now go back to the theory of the brilliant physicist Dr. Freeman Cope. We will combine the magnetic theory (neutral equator) of Davis-Rawls (solid magnets) with the gaseous magnetic theory of Cope. Remember Cope theorized that we live in an atmosphere of *gas magnetoelectric dipoles* (two poles), so our concept must start with both the magnetic field and the electric field thus:

104 Ancient Mysteries, Modern Visions

Magneto (H) Electric (E)

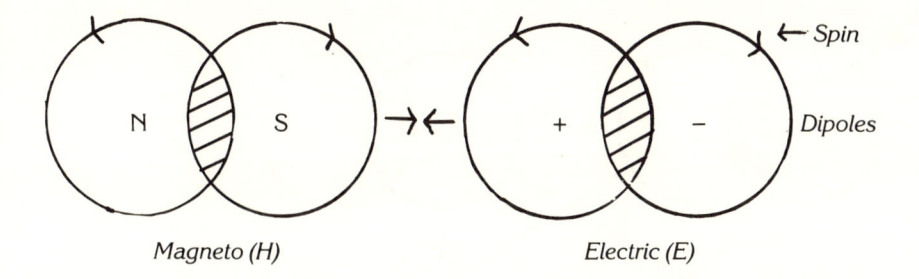

Spin
Dipoles

Since Davis-Rawls used a model where N spins counterclockwise and S clockwise, we will use their spin also. We see then that the south magnetic force would spin *towards* the positive electric force and thus *magnetic south* will bind to electric positive forming a *gas magnetoelectric dipole*. Since conventional radio engineers and scientists use the symbol E for the electric field and H for the magnetic field, I will do the same.

My reader will have noted that I have no rigid orthodoxy against conventional science, amateur scientists or students of the mystic or occult. From my experiences, however, we must listen to all groups. It is inimical to good science that such groups spend their time insulating one another. There is nothing so sanctimonious as a mystic or occult type that thinks he or she is the only one who talks directly to God or the materialistic scientist who thinks he or she *is* God. Most of us, of course, talk to God. It is called prayer. And most of us at one time or another practice science, farmers in particular. The true mystic, of course, is one who practices very hard and becomes good at talking to God, and the true scientist practices hard and becomes good at understanding nature. A few brilliant minds do both, such as the French philosopher-paleontologist Tielhard de Chardin. It is easy to distinguish great mystics or great scientists. They always respect and try to understand one another, and they never badmouth one another—but back to monopoles.

Let us suppose now that these gaseous *magnetoelectric dipoles* are strongly bound together but can be torn apart by high-energy forces, perhaps cosmic rays. We would then end up with two *magnetoelectric monopoles* (one pole). In other words, we would have two HE monopoles floating around in space and each would have a magnetic charge and also electric charge like this:

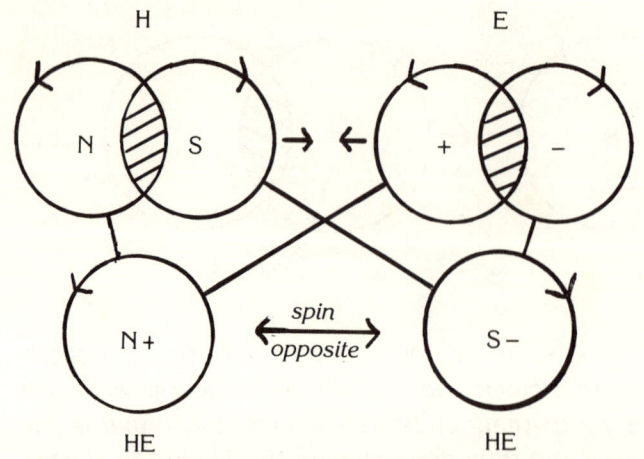

We see then that north and positive go together and south and negative go together. Although forces of the same charge (magnetic and electric) attract, forces of the two different types of charge (magnetic and electric) repel. We therefore end up with *like* forces attracting one another to become HE (magnetoelectric) monopoles. Although Dr. Cope never drew out a model before he died, this is apparently what he meant. We now have two different types of Cope magnetoelectric monopoles floating around, like this:

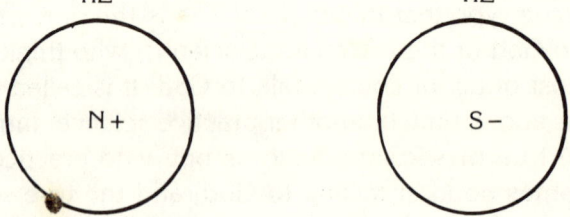

We will now suppose that HE monopoles are weakly bound and under certain conditions, such as sunspot flairs, fly apart like this:

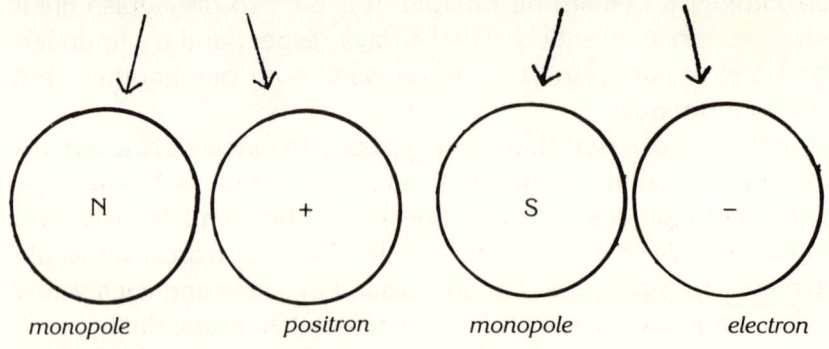

106   Ancient Mysteries, Modern Visions

We see now that during periods of high sunspot activity the weak field that holds the HE monopoles together is bent, so to speak, and torn apart, and what we then have is a massive generation of free, separated fields. The positive and negative free fields are called positrons and electrons (from gaseous ions) and the north and south fields are called north and south magnetic monopoles, also from gaseous ions. Cope's gas magnetoelectric dipoles have split first into magnetoelectric monopoles, then into separate magnetic and electric monopoles.

What has all of this got to do with the para- and diamagnetism that I have studied for so many years in stone, soil and religious structures. That is the simplest part of the model of all. Just as a battery, or a piece of plastic (dielectric), can collect and store positive and negative electrical charges, and north and south magnetic charges, stone and soil can collect and store south HE monopoles and plants north HE monopoles. A piece of plastic such as Teflon that can collect and store electric charges (like an acid battery) is called an electr*et* and is analogous to a magnet, That is why it ends in *et*. Wax is a good electret. I first studied wax electrets in insects as long ago as 1960, and they led to my study of magnets and magnetic monopoles. That is how science really works—one thing leads to another.

We may understand now that a substance that stores south HE monopoles, like granite or porphyry, is called a paramagnetic substance. Volcanic stone is a solid paramagnetic storage battery for south HE monopoles. Oxygen is the most paramagnetic of all gases. It is therefore also a gas storage battery for south HE monopoles. Oxygen, of course, is what keeps man alive and since almost all organic molecules are diamagnetic, man is a paramagnetic, living being only as long as he breathes. Once he dies his body becomes diamagnetic! This is no doubt why the oriental mystics call breath "the spirit of life."

All batteries, whether electric batteries or monopole batteries, have a shelf life—that is the charge trickles off with time. Thus we may understand that over eons of time the stone of the great mountain ranges has been charged with south HE monopoles to become highly paramagnetic, whereas other stone such as the limestone of central Ireland and all plant life (forests) have become charged with north HE monopoles to become diamagnetic. Over

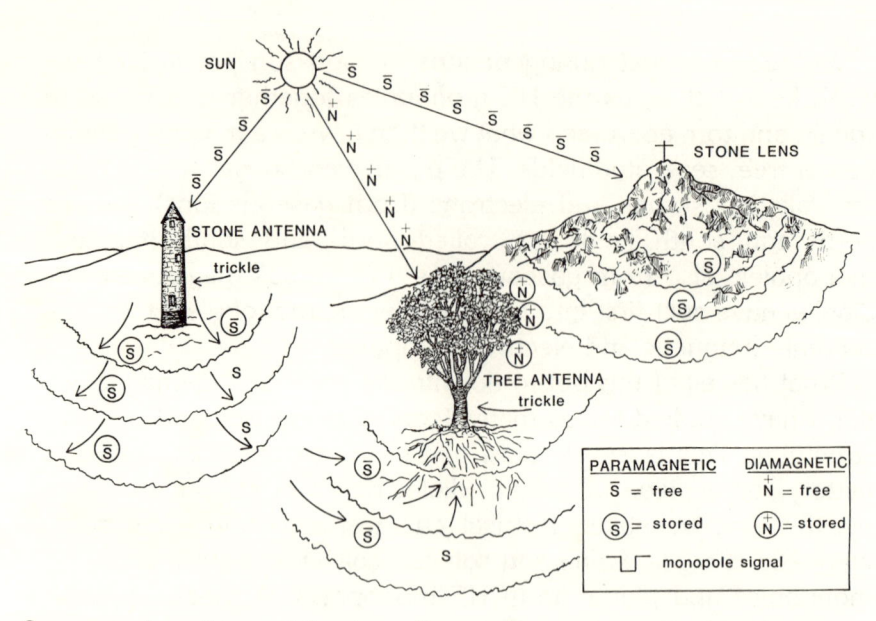

Summary of my theory of the place of magnetic poles in nature. A circle around the monopole means it is stored in the soil, stones or plants. We start at the sun with Cope's magnetoelectric (HE) dipoles (not shown). The dipoles are torn apart by sunflair activity and free magnetoelectric monopoles, south and north, head for earth across space. Some might be absorbed by the atmosphere, but most reach earth where the Ss are being absorbed and stored by stone round towers, stone mountains such as the sacred mountain shown (with cross), and by paramagnetic soil. The N monopoles are adsorbed by plants such as the tree shown. Once adsorbed they are stored like in a battery (circled monopoles). Under certain conditions the stored magnetoelectric monopoles trickle out and separate into south monopoles and electrons and into north monopoles and positrons (separate charges). The free south monopoles in the soil meet with free north monopoles in the plant roots and along with the catalyst of nitrogen, water, etc., set growth and photosynthesis on their way (only the free S monopoles are in in soil, north N monopoles would be in roots).

Sacred mountains and stone religious structures such as round towers and chedi are stone antenna and thus, like radio antennae which collect electron charges, they collect far more south magnetoelectric monopoles than does the soil. They thereby release more monopoles consequently stimulating better growth in crops planted around their base. That is why the Irish monks planted their gardens around the round towers, and the Egyptians planted beautiful gardens around their pyramids and obelisks. The hanging gardens of Babylon were planted on the terraces of a huge steep pyramid. Think also of the crop terracing of China and other Asian countries. Paramagnetic people who are efficient antennae for collecting south magnetoelectric monopoles and release them efficiently as south magnetic monopoles are the healers of the world (or under meditative conditions can levitate). They are sort of human round towers designed by God instead of man.

108 *Ancient Mysteries, Modern Visions*

the eons the volcanic rock erodes into soil that keeps a good charge of south HE monopoles. When a seed or plant is placed in the soil the diamagnetic north HE force meets the paramagnetic south HE force and, with nitrogen as the catalyst, the two force fields break apart into separate charges to promote, in the presence of sunlight, growth and the magic of photosynthesis. They reunite eventually in the plant.

You will notice that although I have emphasized in my work (because they are ignored by agricultural scientists) para- and diamagnetism, I have not practiced *either-or* science. *All* parts of the system are necessary and the sunlight is the generator of those forces.

We see in the figure on the facing page the Callahan theory of magnetic monopoles as it applies to plant growth and thus to agriculture. Magnetoelectric dipoles are torn apart by cosmic forces to form the *free* HE magnetoelectric monopoles of Cope. The weak force that binds these is shattered, by water and soil activity, and N and S magnetic monopoles (free charges) and + and − positrons and electrons (free charges) form in the fertile soil. Finally the free HE charges are collected more efficiently by certain shapes of stone than by random stone forms. In short, the round towers, chedi pyramids and cathedrals of the world are storage antennas for south HE monopoles. Such structures were designed to collect the force and allow it to trickle off more efficiently than it would from other nonsymmetrical forms. The ancients were south HE monopole stone antenna engineers.

There is little doubt that a few certain individuals, for some complex physiological reason, can store and send more free south magnetic monopoles than most people.

These are the healers of the world, such as Father DiOrio (see suggested reading). They belong to many diverse religions, but they all have two things in common—the love of God and their fellow man, and they have been blessed by God with the special physiology that allows them to generate excesses of free south magnetic monopoles. Others like St. Teresa of Avila could generate sufficient magnetic monopoles to actually repel their own bodies or levitate.

Some may ask why south monopoles are the healing-growth force and not north monopoles. That, of course, is like asking why are electrons and not positrons the electronic, radio, TV screen force. I do not know. It is simply, as in all good science, that my experiments fit the model, and so that is *probably* the way God created it.

What I have written is not some crackpot idea. It is based on what other competent scientists have discovered. It is logical because it gives symmetry to the electromagnetic theory. The ancients, who had a brain just like ours (6,000 years is nothing in the theory of evolution), knew how to design stone antennas. Like us they were experimenters.

Whenever I pick up my two volumes by E.A. Wallis Budge on translating hieroglyphics and see those two symbols for stone, I am indeed confounded by the brilliance of our ancestors. Hopefully you will be likewise affected.

Epilogue
The Mummy on the Water

In a marvelous book on Tibet and the Tibetans called *Secret Tibet,* the Italian writer Fosco Maraini recounts his visit with the Tibetan Princess Pema Choki, sister of the Prince Thondup. She describes her girlhood to him and especially the love that she had for her uncle. To quote from that book:

"Pema talked about her uncle again. He was the most extraordinary man I had ever met. I remember that when I was a little girl he lived in a completely empty room and *flew*." [Italics added]

"Weren't you afraid? Did you actually see him?"

"Yes. He did what you call exercises in levitation. I used to take him a little rice. He would be motionless in midair. Every day he rose a little higher. In the end he rose so high that I found it difficult to hand the rice up to him. I was a little girl, and I had to stand on tiptoe . . . there are certain things you don't forget!"

Was Princess Pema Choki lying to Fosco Maraini? Not likely. What would be the point?

Strangely enough the Egyptians left me a sign that they, without a doubt, performed a mummy levitation on their departed priests and pharaohs. Remember a mummy is no longer breathing so it is not only light and dried out, with the organs removed, but highly diamagnetic. Chanting, paramagnetic living priests in a highly paramagnetic stone inner sanctuary (King's Chamber) of the Great Pyramid at Giza, might easily generate enough of Dr. Cope's south HE monopoles to attract upwards a mummy specially prepared with organic chemicals to give off north HE monopoles. Remember north and south HE monopoles spin in opposite directions, and so repel one another.

The trick, of course, would be to generate enough of the two opposite HE monopoles, with their different respective electric and magnetic charges, to fill the chamber before the north and south, positive and negative monopoles flew apart to become free, separate monopole charges. Once that happened the free charges would *float* around until they were absorbed by the granite walls or paramagnetic priests. That, of course, is why performing such a ceremony would make one *feel good.* The priests would be absorbing a lot of energetic south monopoles from the room and walls. This is also why climbing a cliff makes one so strong and energetic. There is a continuous trickle of south monopoles from the paramagnetic stone battery into the living, oxygen-breathing, paramagnetic human body.

What is the sign that the ancient Egyptians left me—a sign that has lasted over 3,500 years? It is the *mummy on the water at Karnak Temple.* I shall never forget the night I first saw it as long as I live.

We had left the hotel in Luxor and taken the bus to the great square in front of Karnak Temple. There was a gentle breeze blowing from the Nile river about a half mile away. As the sun fell behind the horizon I could feel a powerful wind being pulled into the inner courtyard of the great temple. The sun-warmed tapered pylons that guard each side of the temple entrance seem to have been specifically designed to pull cool air into the inner chambers. At the temple entrance the wind force is extremely powerful.

It was our last night at Luxor and we were joining about 500 tourists for the spectacular light show at the Karnak Temple by spotlight illumination. As an Egyptian pharaoh might say, the entire night started out most auspiciously.

Darkness descended, and as the lights shone along the sides of the great entrance pylons, a barn owl, disturbed by the imposition of false daylight, flew out from one of the topmost windows of the south pylon. As I shouted to our group I was struck by the symbolic effect of a pure white bird, flitting like some huge moth back and forth between the pylons and glowing temple pillars. It was almost as if the *soul of the pharaoh had taken flight to heaven!*

Soon the voice of Jack Hawkins, the British actor, blared forth from the hidden speakers. As we 500 awed souls wound our way between the mysterious temple walls and pillars, the history of Karnak was related in the beautiful modulated tones that only the British can impart to English. Five hundred souls by the magic of modern electronics were soon transported back to the torch-lit religious ceremony of Karnak Temple. We were as one in time and space with the pharaoh of Karnak and his priests and subjects. Almost, as somehow without the passing of time, we ended up beyond the sacred lake at that temple dedicated to the Amon.

The Egyptian government provided a huge tier of bleachers along the back of the lake opposite the great east walls of the temple and, as we dutifully filed in to fill the seats, more powerful lights illuminated the great eastern walls of Karnak.

Jack Hawkins' voice drifted out of my mind as I suddenly became aware of another startling phenomenon. I no longer heard the people around me nor even the beautiful accompanying narration. I had grabbed my camera and switched the shutter speed mechanism for a time exposure. I could not see my camera dials in the dark so I held my camera tightly against the bench in front of me (as I had no tripod) and clicked off an entire roll of film—1/8 second, 1/4 second, 1 second, 5 seconds, 10 seconds, etc. I kept track of my exposures by memory and breathed a prayer that at least one photograph would be usable.

Finally the story of Karnak ended and on the way to the bus I asked at least one-third of our group if they had seen it. During the next few days, I asked everyone that I knew had attended the light show if they had witnessed the vision. I didn't get a single affirmative answer! It has been the history of mankind that most individuals may see yet remain blind. The usual question was, "What vision?" "The image of the mummy on the water," I replied.

When I arrived back home in Gainesville, I developed my film and sure enough there was a mummy on the water and that very same image of the levitating mummy would have appeared as thousands of oil lamps lit the temple millenniums ago. The Egyptians had left me a visual image almost 4,000 years old—the

The mummy on the water at Karnak temple, Luxor. This photo should be viewed upside down so that the flood-lighted temple looks like an Egyptian bed with the mummy water reflection in the sacred lake floating above the bed. Why is the entire temple not reflected in the water? The Egyptians were masters of optical science.

message of the levitating mummy on the Sacred Lake of Karnak.

Anyone who has made even a cursory study of the phenomenon of levitation can list hundreds of incidents where perfectly reliable witnesses have observed persons in a state of levitation. All such incidents have at least five things in common.

1. The levitating persons were always holy persons, male or female, who prayed and fasted. They were never fake occult types or sadists (evil). In short, they loved God.

2. The verifiable incidents always took place inside a *stone* building.

3. The holy persons were always in a trance, presumably the low brainwave alpha state (8 cycles per second).

4. They were never aware of the observer entering the room.

5. Most important of all, in most cases the holy mystic levitated against his or her will—that is, they did not plan or instigate the act. This is apparently not true of Yoga or other eastern mystics, such as the uncle of Princess Pema Choki, who *knew* how to levitate.

It would take at least an entire chapter to list all of the saints and holy persons who have been observed by reliable witnesses to levitate while at prayer.

St. Ignatius of Loyola, who died in 1556, was seen by John Pascal to rise more than a foot in the air while in prayer. St. Teresa of Avila not only wrote about her own experiences with levitation, but was often observed in that state by absolutely reliable witnesses. The most famous case concerning levitation was the priest St. Joseph of Cupertino (1603-1663). Apparently the television story of the flying nun was based on the life of Johann (Joseph) Friedrich. The Duke of Brunswick was allowed to hide, with two companions, in the chapel where Father Joseph was observed to *fly*. Many unbiased outsiders observed Joseph as did members of his own order. The *Catholic Encyclopedia* does not duck the issue and says, "There seems to be little doubt concerning the fact of levitation." The reference work lists three possible causes—God, the devil or some force of power of nature as yet unknown.

I do not intend to enter into a philosophical discourse as to whether or not evil (negative occult) can cause levitation. The only reliably witnessed cases I can find in the literature always seem to involve good people, and with good people I include the priests of Egypt. Furthermore, I think it is poor philosophy (which is as ignorant as science) to separate God from nature. I did not say God *is* nature (pantheism), but rather since God created nature He can manipulate it any way He sees fit. If He can *cause* a tiny seed, by the means of the power of electric and magnetic fields, to grow into a corn plant, then most assuredly human levitation, although the exception, is not outside the dynamics of nature.

It is then, in my opinion, and based on what I know about para- and diamagnetism, these forces of nature that are being utilized in levitation, healing, and—most important to my farming friends—growing plants.

It follows that the more we know about these magnetic forces, the more likely we are to involve ourselves in what Charles Walters, Jr. calls *eco-agriculture.* That is, we will not accidentally, through ignorance, deplete from the soils forces that are necessary to plant growth. This seems to me to be far more important than whether mystics can or cannot levitate.

Why then did I even bother to bring the concepts of healing and levitation into my essays on agriculture? Simply because they are fascinating subjects and because they demonstrate that everything really is connected to everything else, consequently the concept of the green thumb becomes a truism, and not some trite saying bandied about by gardeners.

If north and south magnetic monopoles are a good force as I know to be a fact from experimentation with my round towers, and if mystics and healers can generate such an excess of south (paramagnetic) monopoles to levitate or cure, and do so by loving God and man, then those who really love plants must be sending them an excess of the same south magnetic monopoles. In other words, anyone with a green thumb is a person who *loves* plants to the extent that his or her own body *radiates,* so to speak, the plant (being handled) with south magnetic monopoles. *The person and the plant are in magnetic resonance!*

From a purely pragmatic standpoint, however, most farmers simply do not have time to go around *loving* each plant as would a home gardener. From their place in nature, it is far better that the farmer assure that the force comes from the soil on their farm. In other words, that they practice good eco-agriculture. Charles Walters, Jr. puts it in very precise terms in his wonderful book, now out of print, entitled, *The Case for Eco-Agriculture.* The *eco* stands for good *ecology* which is also good *economics* (survival on the farm). To quote Walters:

"The simplicity required by NPK (nitrogen, phosphorus, potassium) sales manuals didn't change biology. Plants still required carbon, hydrogen, and oxygen, and these were still available as carbon dioxide and water—from the air and from soil solutions.

"Plants still required the postively charged (cation) elements—calcium, magnesium, potassium, manganese, iron, zinc, copper and more, available only from the soil, not hydroponic tanks.

"Plants still needed the negatively charged acid elements as anion phases of nitrogen, phosphorus, sulfur, chloride, boron, molybdenum, and so on. College people taking *two anions* and *one cation* by the nape of the neck and the seat of the pants and creating catch phrase NPK didn't exactly change biology. It just made it sound simple, and provided a sort of do-it-yourself brain surgery for the farmer and the oil company salesman.

"How plants are nourished and how they grow is not a complete story, even now. The story had, however, left its primer stage long before the NPK laws made their grand sweep. Swept aside was a term that needs attention again—*the living soil!*"

What Charles Walters, Jr. means, of course, is that the majority of American farmers no longer farm according to the laws of nature (eco-agriculture), but rather according to the *quick fix*. We have become a nation—as has most of civilization, Egypt and Ireland included—of *drug the soil* addicts. Drugging the soil does the same thing for the *living soil* as drugging our bodies does for our living bodies. Not only do the various weed killers and insecticides being poured into the soil by the billions of tons affect the organic content of the soil, but no amount of nitrogen, phosphorus and potassium can replace the rock-generated paramagnetic force that is eroded from our soils by the sheer stupidity of our modern agriculture practices. If my farming friends learn one thing from all of my observations in this book, it is, I would hope, that if their soil does not contain strong paramagnetic force, they will not be able to grow a healthy plant regardless of how much *junk* (I use the word in its modern sense), they pour into the land. Good eco-agriculture leads to healthy plants, and once that state is achieved down on the farm, then insects and disease problems will take care of themselves. Healthy plants, properly nourshed, will fight off attacks from outside, the same way healthy people do.

As I have roamed about between the United States, Ireland and Egypt studying towers, pyramids, rocks and soils, I have become more and more astonished by the simple fact that the same forces that have levitated loving persons, and also that loving persons most often utilized to heal the sick, are the same forces, magnetic and electric, that cause a tiny seed to grow into a corn plant, or even into the giant loblolly pine outside my kitchen door. It is almost as if God has an algebra equation for life:

$$\text{Love} + \text{electromagnetic energy} = \text{life}$$

Unfortunately modern science seems to have forgotten the first term in God's equation.

In the final analysis, whether or not my theories are exact—which they are not as no theory or model ever is—or whether or not my reader believes my every word to be absolutely correct—which also is not indicated, since no one is infallible—what really counts is that every living soul follow the advice that St. Francis of Assisi gave to his friends: "Keep a green bough in your heart and God will send a singing bird."

Appendix 1
How to Make a Corrugated Round Tower Aura Sensor/Soil Tester

The infrared-paramagnetic round tower configuration described here is based on several months' experimentation with different forms and on my experience inside the mysterious corrugated Red Tower of Delhi, the Qutb Minar.

Many species of insects also have corrugated sensilla on their antennae. This is especially true of insects such as mosquitoes that seek out human or mammalian auras. The corrugated sensor is not only more sensitive than the smooth form but also seems to store the aura energy better.

Another reason for describing a corrugated configuration is that it is much easier to accurately taper than a smooth-sided sensor.

1. Fold a five-by-eight-inch index card along the five-inch dimension in one-fourth inch folds. You should end up with approximately fifteen peaks along the eight-inch length. The easiest way to make the folds is to take the card between the fingers of both hands and, using the flat surface of the table as a straightedge, fold the card by pressing it against the table. Turn the card over to alternate the next fold. The tips of your fingers pressed against the fold on the table will keep the corrugations at their quarter-inch dimension.

2. Place a line of white glue all along the outer peak of the last edge; fold at one end and glue it into the inner valley of the last fold at the other end. You now have a round corrugated tower.

3. To taper the tower, take a small rubber band and put it around the top of the tower so that you constrict the upper corrugations. By sliding the rubber band up and down, you can change the degree of taper. If you slide the rubber band to the center of the tower, of course, you will have a smaller diameter (constricted) corrugated tower, but no taper. It is fun to experiment with different degree tapers.

4. In order to hold your index card tower to its shape, put a few drops of glue on the top and also glue the bottom to a little one- or

two-inch square of index card. After a couple of hours, the hardened glue at each end will hold your tapered tower to its shape. Remove the rubber band. You may wish to use your tower without a cone cap. The yellow fever mosquito, which alternately scans human skin with each hind leg, has flat-topped corrugated sensilla on its legs. They look exactly like the corrugated slope without a cone.

5. To make the cone, cut a piece of index card about two inches by three inches in size and fold it along the three-inch edge. Join the edge with a piece of Scotch tape. Cut the corner off to fit the diameter of the top of the tower. Open the cone up and glue it in place. You now have a completed corrugated cardboard round tower.

6. To coat the corrugated tower sensor, spray the index card tower with Scotch Spray Adhesive (photo mount) and sprinkle it with good potting soil, ground limestone, red clay or ground up flowerpot (clay). You may have to spray and coat several times to get a nice, even layer on the corrugated surface. No adhesive spray is needed for a good, wax-tower sensor. Just dip the corrugated tower in melted beeswax.

Suspend the tower from a wooden L-shaped arm by heavy-duty thread. Test your coated tower for paramagnetism with a 1,000-gauss magnet to see if it is strongly attracted—believe me, it will be!

Now that you know how to construct your round tower aura sensor, begin testing it with the paramagnetic-IR aura of your body. The force is strongest at your solar plexus. Test it in a well-ventilated room, but with no breeze blowing. You must first convince yourself that it is a living force in operation and not air blowing the tower toward your body—but then, how does one blow an object toward oneself?

Move slowly, at about a forty-five degree vector, to the side of the front cone of the suspended tower. If you are feeling good—like I do when I'm writing this—the tower will move slowly toward you when your body is four to six inches from it. It always works best at dawn or dusk in a well-ventilated room. Stale air quenches the movement, and people standing too close together jam each other's auras.

If you practice, you will see that your mind affects the movement. It is obviously an alternating or pulsating force since sometimes the tower moves slowly away from your body. The tower definitely stores energy, as you will soon note that there is a considerable pause before it moves to your body.

The beauty of this round tower aura sensor is that unlike a lot of the occult systems, it always works! One does not need to become an expert yogi to draw the tower to the body. You will find that a tower that you make always works better for yourself than for your friends, although it will respond to anyone. Why does it work best for you? Obviously, because in constructing it, you have "doped" it with the "essence of yourself!"

Finally, of course, you can use round towers such as these for soil testing by coating them with different types of soil and using a protractor to measure the distance the various soil-coated towers rotate in a fixed, 1,000-gauss magnet. The stronger the attraction, the more paramagnetic the soil, and hence, the better the soil.

In these days of modern stress when the world seems filled with vicious fanatics whose solution to world problems is to kill, how stimulating of mind it is to celebrate life by testing your own living aura and, likewise, to celebrate the good earth by testing soil for that mysterious, infrared-paramagnetic earth force.

Good luck and have fun!

Appendix 2
Technique for Plotting Paramagnetic Force Fields

The paramagnetic force field on model carborundum round towers is easily demonstrated by a simple technique. Figure 1 shows a close-up of the field of paramagnetic energy on a round tower made of 183W (T461) *Turbac Durite* carborundum sandpaper. Heavyweight Durite is used for polishing metal and is covered with a thin coating of plastic to protect it from the cooling water.

The round tower is submerged in a jar with 12 tablespoons of Epsom salts dissolved in a quart of water. Leave the tower in this solution approximately one day. After a day it is removed and set out to dry naturally. The paramagnetic carborundum orientates the diamagnetic Epsom salt crystals into straight "lay lines" of energy.

It will be noted from Figure 1 that the force lines are almost exactly one millimeter apart (see ruler to right of photograph).

Anywhere there is an imperfection (crack or fold) in the smooth surface of the plastic-coated carborundum sheet the Epsom salt

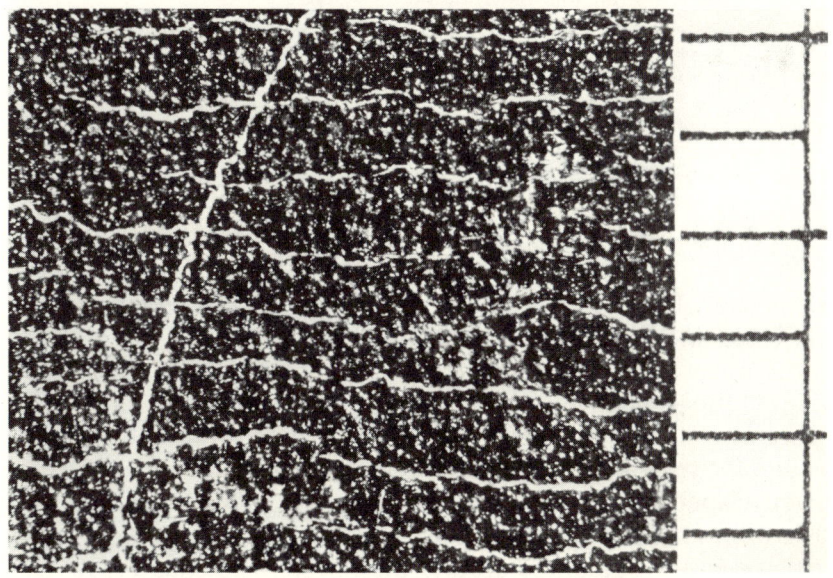

Figure 1. Force lines of Epsom salts on carborundum tower. Note that they are spaced evenly one millimeter. Microphotograph by Pat Greany.

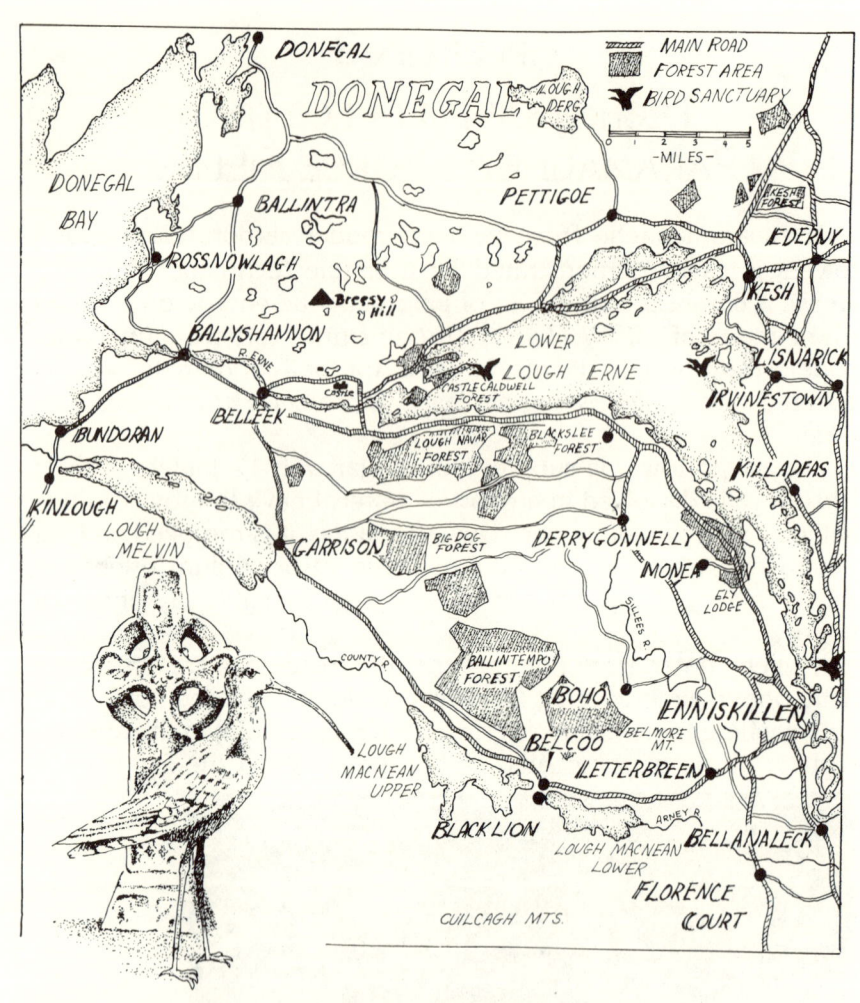

Figure 2. Map of area around Belleek, County Fermanagh, Ireland.

line is likely to orientate in different directions as is shown by the one cross line in the photograph.

It is observed that the lines always orient across the nine-inch width of the 9-by-11-inch sheet of paper, and not the length. Thus if the tower is formed by rolling the sheet lengthwise the lines go around the tower, and if rolled width-wise they go up and down the tower. Experiments presently are being run to test the effect of towers with horizontal force fields versus towers with vertical force fields on radishes planted around the towers (see chapter 3).

This same technique can be utilized to plot the force fields on a map of a geographical region. Figure 2 is a map of the region

122 *Ancient Mysteries, Modern Visions*

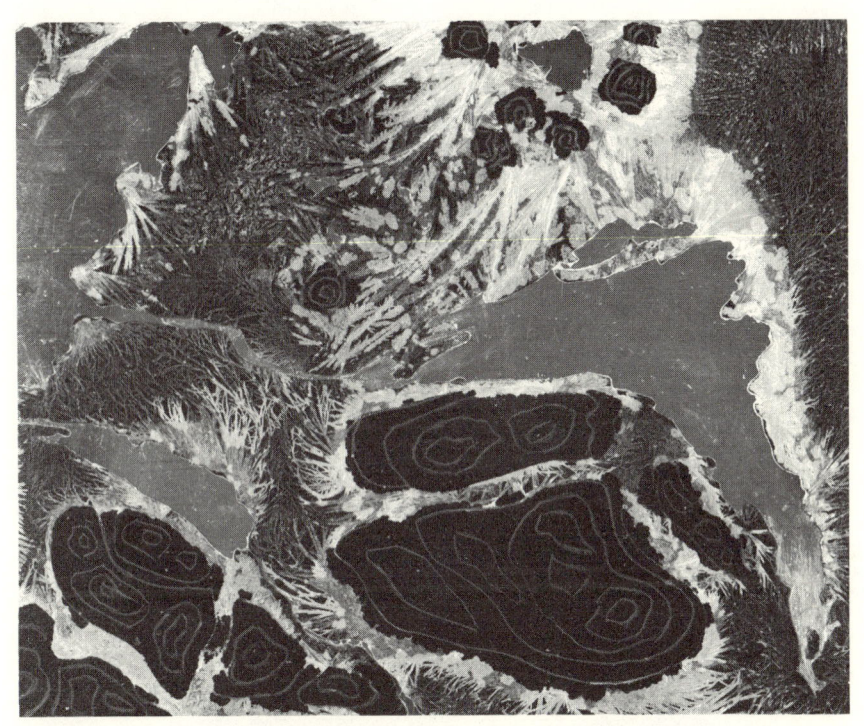

Figure 3. Paramagnetic carborundum force field map of area in figure 2 around Belleek.

around Belleek County, Fermanagh, Ireland. It is the center of the "healthy area" described in chapter 1. Keenaghan Lough is shown across the road from the castle at the west end of Lough Erne; Breesy Hill lies just north. The River Erne flows from Lough Erne (right) to Donegal Bay (left) past Ballyshannon. Breesy Hill is the *magic* mountain where I often went to revive my spirits during World War II.

Utilizing an Irish Survey Map (1 inch per mile), I carefully traced and indented with a metal stylet the contours of the map on a sheet of *183W Turbac Durite* carborundum (Figure 3). The sandpaper was mounted on a sheet of aluminum metal. The lakes, River Erne, and Donegal Bay were cut out of the sandpaper pattern leaving the aluminum (light grey) as water.

With a tablespoon I carefully flooded the valley and lakeshore areas *between the mountain contours* with saturated diamagnetic Epsom salts.

The carborundum map was allowed to dry naturally for two days and then photographed.

Appendix   123

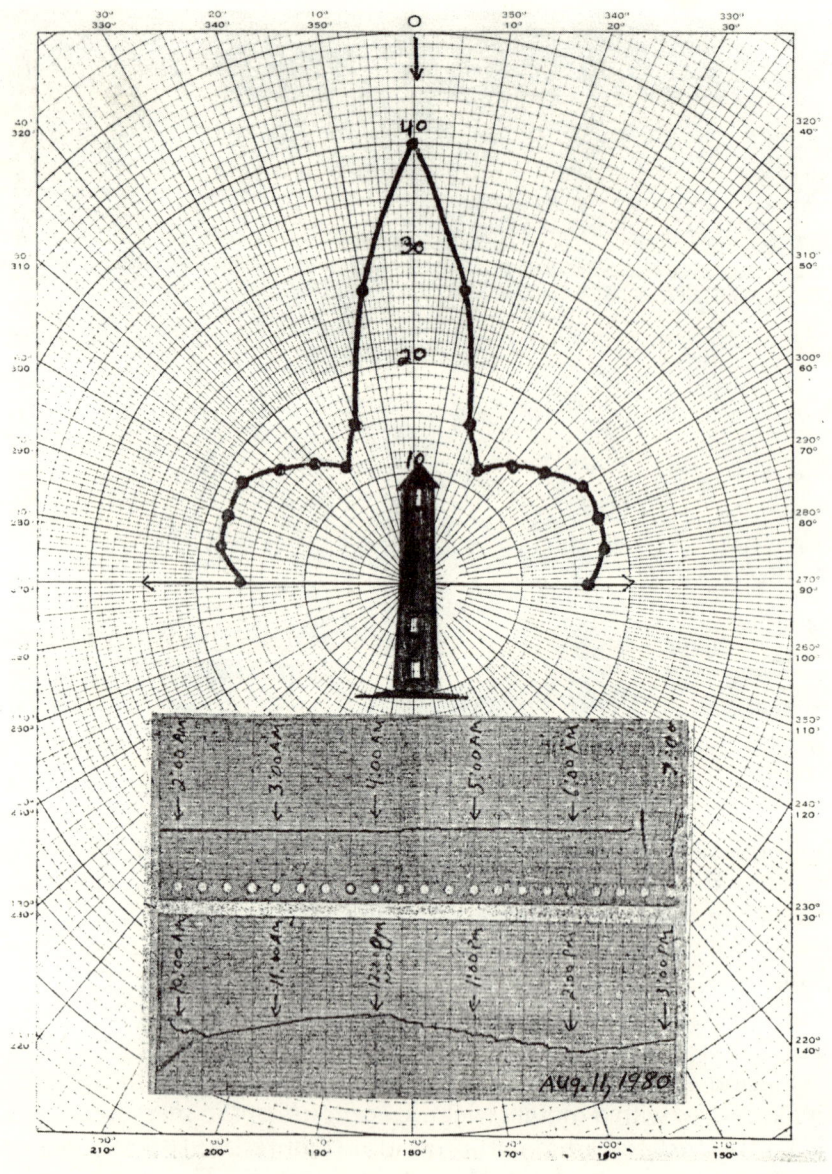

A round tower model is rotated 360 degrees in a centimeter radio wave field and the maximum energy is plotted on polar paper. Radio engineers would call a round tower an "end fire array" since most of the amplified radio energy is guided down the shaft from directly above. Note that when it is rotated 80 to 90 degrees smaller sidelobes of energy appear. Such patterns are common for radio antennae. Below, electrometer recording of detected waves. The top recording is night time and is flat, bottom is daylight, curved, when the sun is on the tower. The sun emits 3-10-centimeter waves.

124   *Ancient Mysteries, Modern Visions*

It will be noted that the sacred Lough Derg, where St. Patrick is believed to have meditated (figure 3, top right center), and its surrounding hills (seven clear, contoured mountains) is almost white with the force lines, and that from this area long lines of paramagnetic force sweep south to Breesy Hill (lone contoured mountain). They concentrate right at Keenaghan Lake where the ancient Celtic Abbey and healthy area lies. Note also that the north edge of Lough Erne is highly paramagnetic. It is in these areas where all of the ancient megalithic structures and Celtic stone structures have been found. There is also a monastery at Rossnowlaugh on the coast of Donegal Bay. Strong force lines sweep into that beautiful region of Donegal Coast.

The implications of this type of paramagnetic (template) force field map to the works of map dowsers should be obvious to even the most critical of map dowsing, especially when one considers that paper is *diamagnetic* and most inks are highly *paramagnetic!*

Paramagnetism and diamagnetism are extremely weak forces, but over a period of time it is quite simple to demonstrate their *slowly* developing steady power. This is of course the way *life* really works—not by *"hitting everything over the head with a high energy inorganic hammer."* The rock-organic cosmic unity of life is slow and steady-but sure. It is a detriment to our modern technology that present day science is so narrow and impatient as to neglect the study of these weak accumulative forces of nature.

Acknowledgement

The most important chapter in this book is chapter 7, *The Detection of Magnetic Monopoles and Tachyons — A Picture of God*. Several wonderful persons were responsible for educating me on the subject of monopoles and tachyons and in giving me the courage to *publish* this chapter.

Dr. Freeman Cope, a physicist now deceased, spent many hours on the phone from the Naval Air Development Center in Pennsylvania explaining to me his elegant theory on tachyon magnetoelectric dipoles. My good friend David Pace, a mathematician at the University of Florida, led me through the maze of complex formulas (the path is not always obvious to a biologist) while he was working in my USDA laboratory.

Finally, I would like to thank Andrija Puharich, M.D., a physicist, and Dr. Elizabeth A. Rauscher, professor of physics at the Lawrence Berkley Laboratory at the University of California, for their work in an elegant monograph titled "The Iceland Papers — Selected Papers on Experimental and Theoretical Research on the Physics of Consciousness" (1979).

At a delightful meeting in Madrid, Spain in 1980, Dr. Puharich, who edited the papers of the Iceland meeting, gave me a copy of that volume. It was my first real introduction into the never-never land of tachyons, although I had been aware of them and began me plant experiments in 1979.

Finally, of course, I would like to thank Charles Walters, Jr. for having the faith in me to publish herein this first possible experimental detection of tachyons.

Whether I am right or wrong, only time will tell, however, I *delight* in the simple fact that certain skeptical theoretical physicists are going to have a very difficult time trying to *explain away their own exact theoretical and mathematical predictions,* predictions that my *precise* plant recording fulfilled.

Annotated Bibliography

IRELAND

The book by Barrow on *The Round Towers of Ireland* is an excellent gazetteer and source of information on round towers. *The Celts* by Herm is a detailed history of those fascinating people. By far the best book on Ireland's old agricultural ways is Evan's *Irish Heritage*. Mitchell's *The Irish Landscape* is the best geological work I have ever read on a single country.

Baille, I.F. and S.J. Sheehy, Editors. *Irish Agriculture in a Changing World.* Edinburgh: University of New Castle upon Tyne and Oliver and Boyd, 1971.
Barrow, George Lennox. *The Round Towers of Ireland.* Ireland: The Academy Press.
de Paor, Maire and Liam. *Early Christian Ireland.* New York: Frederick A. Praeger, 1958.
Evans, E. Estyn. *Irish Heritage, The Landscape, The People and Their Work.* 9th edition. Dundalk, Ireland: W. Tempest, Dundalgan Press, 1967.
Evans, E. Estyn. *Irish Folk Ways.* 5th edition. London and Boston: Routledge and Kegan Paul, 1972.
Herm, Gerhard. *The Celts, The People Who Came out of the Darkness.* London: Weidenfeld and Nicolson.
Mitchell, Frank. *The Irish Landscape.* London: Collins, 1976.
Pochin Mould, D.D.C. *The Mountains of Ireland.* Ireland: Gill and Macmillan, 1976.
Scherman, Katherine. *The Flowering of Ireland, Saints, Scholars and Kings.* Boston and Toronto: Little Brown and Company, 1981.

EGYPT

For an excellent overview of Egypt and her history give a close look at *Atlas of Ancient Egypt,* by Baines and Malek. If you wish to learn hieroglyphics you will certainly need Budges two dictionary volumes. For a beginner the little pocket size *Egyptian Language* by Budge will give all that is needed to understand hieroglyphics. *The Nile* is a masterpiece by the great German archaeologist, Emil Ludwig. For an overview of the pyramids read Edwards, *The*

Pyramids of Egypt. The best general summary on ancient Egyptian history and culture is *Ancient Egypt* by White.

Baines, John and Jaromir Malek. *Atlas of Ancient Egypt.* New York: Facts on File Publications, 1980.

Brier, Bob. *Ancient Egypt Magic.* New York: William Morrow and Company, 1980.

Budge, E.A. Wallis. *Osiris and the Egyptian Resurrection.* Reprint of 1911 work. New York: Dover Publications, 1973.

Budge, E.A. Wallis. *Egyptian Hieroglyphics Dictionary.* Two volumes. Reprint of 1920 works. New York: Dover Publications, 1978.

Budge, E.A. Wallis. *Egyptian Language, Easy Lessons in Egyptian Hieroglyphics.* 16th edition. New York: Dover Publications, 1978.

Clark, R.T. Rundle. *Myth and Symbol in Ancient Egypt.* London: Thames and Hudson, 1959.

Clayton, Peter A. *The Rediscovery of Ancient Egypt, Artists and Travelers in the 19th Century.* New York: Thames and Hudson, 1982.

Edwards, I.E.S. *The Pyramids of Egypt.* 16th edition. Middlesex, England, New York: Penguin Books, 1979.

Erman, Adolf. *Life in Ancient Egypt.* Reprint of 1894 work. New York: Dover Publications, 1971.

Fagan, Brian M. *The Rape of the Nile, Tomb Robbers, Tourists, and Archaeologists in Egypt.* London: Macdonalds and Janes, 1977.

Frith, Francis. *Egypt and the Holy Land in Historic Photographs,* introduction by Julia Van Haaften. Reprint selection from 1862 volumes. New York: Dover Publications, 1980.

Ludwig, Emil. *The Nile, The Life Story of a River,* translated by Mary H. Lindsay. New York: The Viking Press, 1937.

Tompkins, Peter. *Secret of the Great Pyramid.* New York, London: Harper and Row Publications, 1971.

White, J.E. Manship. *Ancient Egypt, Its Cultures and History,* reprint of a 1952 work. New York: Dover Publications, 1970.

ANCIENT MYSTERIES

For a unique theory on the ancient connection between Ireland and Egypt read Ivimy's *The Sphinx and the Megaliths.* The author puts forth the idea that Stonehenge and other stone rings in Ireland

and England were constructed by an Egyptian colony established by priests of the sun god Re. His theory is not without merit. In *America B.C.* the Harvard professor Barry Fell makes a good case for peoples settling in America, before Christ, from European regions, especially Celtic Spain, Carthage, Libya and Egypt. Artifacts from European cultures have been found all over North America from New England to Oklahoma. A good overview on the life of a modern healer is Father Di Orio's book, *Called to Heal.* The books by Schul and Pettit and Pat Flanagan make a good case for a pyramid power still in existence. Smith's, *The Image of Guadalupe* is a fascinating story of the miracle of the Image of Guadalupe. *Ley Lines* is the best on that subject.

Abehsera, Michel. *The Healing Clay.* Brooklyn, New York: Boulder Books-Swan House, 1977.

Callahan, Philip S. *The Tilma Under Infrared Radiation.* Cara volume II, number 3. Washington, D.C.: 1981.

Di Orio, Ralph A. *Called to Heal.* Garden City, New York: Doubleday and Company, 1982.

Fell, Barry. *America B.C.* New York: Pocket Books, 1976.

Fell, Barry. *Saga America.* New York: Times Books, 1980.

Fidler, J.H. *Ley Lines.* Wellingborough, England: Turnstone Press, 1983.

Flanagan, Pat G. *Pyramid Power.* Marina del Rey, California: De Vorss and Company, 1973.

Frazier, James George. *The Golden Bough.* New York: Macmillan Publishers, 1922.

Ivimy, John. *The Sphinx and the Megalith.* London: Abacus, 1976.

Maraini, Fosco. *Secret Tibet.* New York: Viking Press, 1953.

Mendelssohn, Kurt. *The Riddle of the Pyramids.* London: Sphere Books Limited, 1974.

Rogo, D. Scott. *Miracles—A Parascientific Inquiry into Wondrous Phenomenon.* New York: The Dial Press, 1982.

Schul, Bill and Ed Pettit. *The Secret Power of Pyramids.* London: Coronet Books, 1975.

Smith, Jody Brant. *The Image of Guadalupe.* Garden City, New York: Doubleday and Company, Inc., 1983.

Toth, Max and Greg Nielsen. *Pyramid Power.* New York: Warner Destiny Books.

GEOLOGY

Rocks and Minerals by Pearl is one of the best small handbooks for gaining a very basic understanding of rocks and minerals. For identification the Peterson series field guide, *A Field Guide to Rocks and Minerals*, by Pough, will acquaint the beginner with most common minerals and stones. For good summary of how certain land forms evolve read *Land from the Sea* by Hoffmeister. It is about the formation of South Florida. Binion's *Scenic and Historic Landmarks* has a marvelous description of Hueco Tanks near El Paso, Texas.

Binion, Charles H. *An Introduction to El Paso's Scenic and Historic Landmarks.* El Paso, Texas: Texas Western Press, The University of Texas, 1970.

Dana, James D. *Dana's Manual of Mineralogy,* 17th edition, revised by C.S. Hurlbut, Jr. New York: John Wiley and Sons, 1966.

Hoffmeister, John H. *Land from the Sea, The Geologic Study of South Florida.* Coral Gables, Florida: University of Miami Press, 1974.

Pearl, Richard M. *Rock and Minerals.* New York: Harper and Row Publishers, 1956.

Rough, Frederick H. *A Field Guide to Rocks and Minerals,* Peterson Field Guide Series. Boston: Houghton Miffline Company, 1976.

MAGNETISM

Although there is only one small reference to para and diamagnetism in *Paleomagnetism* by Irving, nevertheless it is the most detailed text on magnetism in geological formations. The books by Davis and Rawles, although ignored by conventional scientists, are classics in their own right.

Davis, Albert R. and Walter C. Rawls, Jr. *Magnetism and Its Effects on the Living System.* Smithtown, New York: Exposition Press, 1980.

Davis, Albert R. and Walter C. Rawls, Jr. *The Magnetic Effect.* Smithtown, New York: Exposition Press, 1980.

Irving, E. *Paleomagnetism and Its Application to Geological Problems.* New York: John Wiley and Sons, 1964.

THE NIGHT SKY

The Stars, by Rey, is without question the most understandable words on the complexities of the heavens that I have ever read. Other than my work, it is the only treatise on astronomy I know about that, like myself, suggests that the Great Pyramid was lined up on Thuban, the former pole star.

Moor, Patrick. *The Pocket Guide to Astronomy.* New York: Simon and Schuster, 1980.
Rey, H.A. *The Stars.* Boston: Houghton Mifflin Company, 1976.

AGRICULTURE

In this section are listed three classic works on the art and practice of eco-agriculture.

Walters, Charles, Jr. and C.J. Fenzau. *An Acres U.S.A. Primer.* Acres U.S.A. Kansas City, Missouri, 1979.
Walters, Charles, Jr. *The Case for Eco-agriculture.* Acres U.S.A. Kansas City, Missouri, 1975.
Willis, Harold. *The Rest of the Story—About Agriculture Today.* Published by author, P.O. Box 692, Wisconsin Dells, Wisconsin, 1983.

THE NEW PHYSICS

This section lists books that give scientific insight into a *New Physics* that marries science, philosophy and spirituality.

Callahan, Philip S. *Tuning In To Nature, Solar Energy, Infrared Radiation* and the *Insect Communication System.* Old Greenwich, Connecticut: Devin-Adair Company, 1975.
Callahan, Philip S. *The Soul of the Ghost Moth.* Old Greenwich, Connecticut: Devin-Adair Company, 1981.
Capra, Fritjof. *The Tao of Physics.* Boulder, Colorado: Shambhala Publications, 1975.
Zukav, Gary. *The Dancing Wu Li Masters.* New York: Bantam New Age Books, 1979.

INDEX

A

Abyssinia, 93
Abyssinian, highlands, 92, 93, 94
Abyssinian, floods, 94
AC motor, 17
Acres U.S.A., 35
Acres U.S.A. Primer, An, 2
Acid battery, 107
Africa, 54, 92
Air vent, King's Chamber, 60
Alcohol, 69
Alga, Pithophoro, 34
Alluvium, 93
Alpha state, 114
Aluminum, fillings, 30
Amateur, 105
America, 73
Ancients, 72, 91, 109
Aner, 98
Anglo-Irish, 10
Animal magnetism, 46, 48
Anions, 116
Ankh, 89, 92, 93
Antigen-antibody, 75
Antigravity, 30, 50
Anti-submarine, 14
Antrim, 12
Antenna, 15, 21, 22
Antenna, coil, 20
Antenna, detector, 76
Antenna, dielectric, 58, 76
Antenna, engineers, 109
Antenna, focusing round tower, 36
Antenna, nature, 36
Antenna, insects, 42
Antenna, paramagnetic, 56
Antenna, paper wasp, 57
Antenna, plant, 78
Antenna, radio, 108
Antenna, radio-optical, 58
Antenna, solar, 70
Antenna, stone, 64, 71, 108, 110
Antenna, stone and soil, 42, 72
Antenna, superconductor, 82
Antenna, waveguide, 24
Arboe, 12
Archaeologists, 54, 85
Ardrahan, 7
Argillaceous, clay, 28
Armagh, 9, 13
Asia, 42, 108
Astrologers, 20
Astrology, 18
Astronomers, round tower, 1
Aswan, 62, 63, 89, 95
Aswan Dam, 54, 89, 93
Atbara River, 93
Atom, 100
Atomic orbitals, 23, 27
Atoms, of wood, 43
Attraction, plant, 70
Aura, colored clouds, 75, 76
Aura, human, 61
Aura, infrared, 49
Aura, sensor, 51, 60
Aztec, language, 50
Aztec, pyramid, 50

B

Babar Archipelago, 68
Bank of dead, 87
Bank of living, 87
Barn owl, 112
Barnothy, M.F., 34
Barrow, G.L., 4, 6, 16, 22, 24
Bartholomeu world map, 88
Basalt, 27, 70
Basilica of Guadalupe, 100, 101
Battery, 107
Beans, 35
Beeswax, 46, 59
Beliefs, Christian, 91
Bell Laboratory, 21
Bell Telephone, 46
Bell towers, 3, 26
Belleek, Ireland, 3,
Bird of prey, *xv*, 61
Big Dipper, 66
Biology, 116
Biome, 2
Blue Nile, 89, 92, 93
Boat, radio-controlled, 17
Boe, A., 35
Boeing 747, 102
Bohr, Niel, 102
Book of the Dead, 91, 95
Boron, 116
Breast and arm, 97
Breath, 50, 64
Breath, paramagnetic, 99
Breath, spirit of life, 107
Breesy Mountain, 15
British Museum, 13
Brown, Jr., F.R.A., 35
Brown, Joe, 69
Buddha, 42

132   *Ancient Mysteries, Modern Visions*

Buddhist monasteries, 42
Budge, Wallis, E.A., 85, 91, 97, 110
Burbank, Luther, 25
Burma, 42, 66
Burren, Ireland, 7, 10
Butterfield Stage, xv, xix

C

Cabbage looper moth, 47
Cabrera, Dr. Blas, 74, 77
Cactus, 100
Caesar, 1, 2
Cairns, 68
Cairo, 38, 88, 92, 95
Calcite, 27, 58
Calcium, 116
Caledonia, upheavel, 12
Camelopardalis, 7, 9
Camera, 113
Canada, 35
Candle, 60
Canyon de la Virgin, xiv, xix
Canyon of Leguma Priety, xvi
Carbon dioxide, 116
Carbon, electrode, 23
Carbonate, 28
Carborundum, 21, 22, 27, 28, 29, 32
Carn, Ireland, parish of, 15
Carpathian Mountains, 2
Carver, George Washington, 25
Casing, of pyramid, 59
Cassiopeia, 7, 9, 21
Cathedral, Egyptian, 40
Cathedral, Gothic, 40
Cathedrals, 109
Cation, 116
Catholic, 64
Catholic Encyclopedia, 115
Cat's whisker, 21
Cave of the Lion, xvi, xix
Celtic ancestors, 16
Celtic defensive structures, 6
Celtic expansion, 1
Celtic friars, 36
Celtic monks, 15, 26
Celtic sites, 14
Celtic warriors, 2
Celts, 1, 2, 3, 13, 15, 18, 21, 98
Centimeter, 19, 21, 23
Central America, 50
Ceremonial rites, 68
Champollion, le Jeune, 97
Chang Mai, Thailand, 44
Chanting priests, 50
Charcoal, xviii
Charge, magnetic, 105
Charge, monopole, 112
Charge, negative and positive, 107

Chedi, 40-44, 48, 50, 57, 108, 109
Childbirth, 68
China, 56, 108
Chinese, 56
Chloride, 116
Choki, Pema, Princess, 111, 114
Christ, 92
Christ, blood of, 64
Christian, American, 40
Christian, immortality, 85
Christian, tourist, 92
Christianity, 13
Cigarette smoke, 23
Circle Gardening, 56
Civilization, Egyptian, 89, 92, 95
Clare, County, 7
Clark, Kenneth, 40
Clay, healing, 55
Clay, minerals, 58
Clay, paramagnetic, 54
Clay, slate, 27
Cleaning, housefly, 48
Cliffs, rock, 70
Climbers, high, 69
Clones, 12
Clonmacnoise, 7, 9, 13
Cobalt, 30
Cobalt, magnet, 95
Coherent waves, 15
Coke, 23
Comanches, xv
Computer, plant, 79, 80, 82
Condensing lens, 50, 59
Cone, 28
Cone, shape, 71
Cone, sensor, 47
Crops, 95, 108
Crystal set, 22
Constellations, 8, 9, 12
Continuous wave, 79
Cope, Freeman, 72, 74-78, 81, 82, 104, 106, 107, 109
Copper, 116
Copper oxide, xviii
Corben, H.C., 78
Corn earworm moth, 42, 45
Corundum, 28
Cosmic energy, 50, 64
Cosmic force, 109
Cosmic origin, 72
Cosmic rays, 105
Cosmic sky, 25
Cosmos, 19, 21
Cotter, house, 98
Cotton, 53
Cotton, crop, 92
Cotton, plant, 80
Crab Nebula, 20
Creationism, 102

Index 133

Crocodile god, 88
Crop growth, 73
Crops, 94
Cross and circle, 92
Crystal detector, 21
Crystal radio, 20
Cycle of life, 89, 92
Cyst, 55

D

Dana's Manual of Mineralogy, 28
Davis, A.R., 34, 103
Day of infamy, *xx*
DC, rectifier, 20, 21, 22
DDT, 62
Death, 91
Decibel, 23, 24
Denderah, 13
Detector, monopole, 81
Detector, plant, 80
Detector, rectifier, 20
Determinative, 99
Devenish Island, 3, 23
Devenish Round Tower, 4, 9, 12, 14, 21, 29, 31, 34
Devil, 115
Diamagnetic, 23, 30, 36, 46
Diamagnetic, bedrock, 98
Diamagnetic, HE force, 109
Diamagnetic, plants, 42
Diamagnetic, trees, 43
Diamagnetic, water, 89
Diamagnetism, 94, 95, 100, 107
Dielectric, 22, 23, 57
Dinder River, 93
DiOrio, Father Ralph A., 109
Dipole, 72
Diptera, 43, 57
Dirac, P.A.M., 72, 75, 77
Directional studies, 21
Dirt filled space, 24, 33
Dispersal, behavior, *xvi*
Divining, 46, 72
Dolomite, 58
Doorway, 24, 33
Doped, blood, 50
Doped, limestone, 58
Doped, oak, 43
Doping, 23, 36, 46, 61
Doping, body, 69
Dowsing, 46, 72
Draco, the dragon, 7-9, 12-14
Dragoman, 38, 39
Dragon, Chinese, 13
Drain, line, 12
Dreenan, cairn, 14
Drug, soil, 117
Drumcliff Round Tower, 7, 12

Drumlane Round Tower, 12
Dublin, 6
Duke of Brunswick, 115
Dunce hat, 27, 29
Dunlop, D.W., 34
Dysert O'Dea Round Tower, 7

E

Earth spin, 13
Earth spots, *xx*
East bank, Nile, 87
Ecclesiastical center, 9
Ecliptic center, 13
Ecliptic pole, 8, 9
Eco-agriculture, 2, 115
Eco-Agriculture, The Case For, 116
Edfu, 88, 95
Effects, biological, 34
Effects, of magnetism, 34
Eggshell thinning, 62
Egypt, 13, 40, 53-55, 90, 117
Egyptians, 8, 13, 14, 18, 40, 87, 96-98 108, 111, 112
Egyptians, ancient, 100
Egyptians, bed, 114
Egyptians, beliefs, 91
Egyptologist, 92, 97
Egyptologists, Danish, 88
Einstein, Albert, 30, 67, 100
"Either-or" science, 109
Electret, 107
Electric field, 77, 105
Electrical charges, 73
Electromagnetic energy, 22, 27, 117
Electromagnetic Fields and Life, 35
Electromagnetic radiation, 19
Electromagnetic theory, 110
Electromagnetic waves, 20
Electrometer, 15, 19, 24, 76
Electrometer, Keithley, 77, 78
Electron, charge, 72
Electron detector, 75
Electrons, 74, 77, 100, 107, 109, 110
Emery, hematite, 28
Emery paper, 29
Encyclopedia, 22
Energies, weak, 19
Energy, collector, 57
Energy, high, 92
Energy, paramagnetic, 91
Energy, rock, 70
English, 12
Entomologist, 42, 48, 73, 76, 82
Epsom salt, 30, 31, 32
Equator, of magnet, 103
Eroded, stone, 95
Erosion, 93
Essence of iron, 43

134   Ancient Mysteries, Modern Visions

Essene Gospel, 55
Eta Draconis, 8, 9, 12, 14
Evolution, 110
Exoskeleton, of insect, 46

F

Falcon, 69
Falcon, American kestrel, 62
Falcon, European kestrel, 62
Falcon god, 64, 89
Falcon, kestrel, 40, 42, 50, 61, 64
Falcon, lesser kestrel, 65
Falcon, mummified, 40
Falcon, peregrine, 42
Falcon, saker, 40, 42
Falconer, 69
Famine, 87
Faraday, Michael, 42
Farmers, 73, 98, 102, 105, 116, 117
Farming, 34, 115
Farming, holistic view, 48
Fatigue, 69
Fatigue transferal, 68, 69, 70
Feather, 99
Fenzau, C.J., 2
Fermanagh, County, 23
Ferromagnetic, 28
Fertility, 68, 91
Fertility charm, 68
Fertilizer, manure, 2
Feynman, R.P., 102
Feynman diagram, 102
Ficus benjamina, 76-79, 81, 82, 102
Field, earth's magnetic, 103
Field, electric, 104
Field lines, 31
Field, magnetic, 42, 104
Field theory, 30
Fields, electric and magnetic, 115
Fields, free, 107
Fields, magnetic, 36
First people, 14
Floods, 94, 95
Floors, lens-like, 41
Floors, of towers, 32, 59
Flowerpot, 27-29, 36, 55, 70, 71
Flowerpot, absorption, 30
Flowing water, magnetic charge, 72
Fluorescent, growlight bulbs, 70
Fluorescent tube, 17
Force, cosmic, 59
Force, cosmic, paramagnetic, 58
Force, fixed, 59
Force, magic, in clay and stone, 54
Force, magnetic, 69, 103
Force, magnetic and electric, 117
Force, of lines, 32, 35
Force, of nature, 115

Force, paramagnetic, 50
Force, paramagnetic, rock, 117
Force, weak, 59
Force field lines, 99
Force lines, paramagnetic, 98
Forces, earth, 90
Forces, high-energy, 105
Forces, like, 106
Forces, magnetic and electric, 106
Forces, physical, 94
Forces, weak, 109
Forces, weak, natural, 94
Formaldehyde, mummify, 40
Form, body, for psychokinesis, 48
Form, paramagnetism, 71
Forms, religious, 40
Fourier transform, 80
Franklin, Ben, 17
Franklin Range, El Paso, *xx*
Fraizer, Sir James George, 68
Frederick II, emperor, *xv*

G

G plot, 84
Gainesville, Florida, 61, 113
Galena, 21
Gallachair, Reverend P.O., 14, 15
Gallagher, Thomas, 14
Galvanometer coil, 79
Galvonometer needle, 78
Galway, County, 12, 33
Garden paradise, 96
Gardens, 108
Gaul, 1
Gauss, 28
Gawaad, Doctor Abdel, 53
Geobiologists, German, 15
Geopathogenic, 15
Germination, 36
Giant particles, monopole, 77
Giza, 38, 39, 59
Glass lens, 23
Glendalough, 7, 11
God, 100, 102, 105, 109, 110, 115
God, love, 116
God's equation, 117
God's handiwork, 84
Godel, Esker, Bach, an Eternal Golden Braid, 84
Golden Bough, The, 68
Gothic cathedral, 57
Graffiti, *xix, xx*
Grand Unified Theory, 77
Granite, 27, 70, 99, 107
Granite, pink, 41, 59, 64
Granite walls, 112
Granules, susceptible versus nonsusceptible, 70

Graveyard, Keeneghan, 15
Graveyard site, 14
Gravitational pull, 19
Gravity, 25
Greenery, 92
Green thumb, 115, 116
Grid lines, 76
Guadalupe River, San Antonio, *xx*
Guerilla warfare, 1

H

H field, 105
Hangchow, China, 43, 56
Hanging gardens of Babylon, 108
Hard Years, The, 69
Harmonious points, 15
Harvester, wheeled, 2
Hawkins, Jack, 113
Hawks, 62
HE magnetoelectric dipols, free, 108
HE monopole, magnetoelectric, 105, 106, 107, 109, 112
Healers, 108, 116
Healing, 68, 115
Healing/growth force, 109
Healthy plants, 117
Heat, as mode of motion, 43
Heaven, hell, 91
Hebrew, 92
Hematite, 28
Hern, Gerhard, 1
Heron, *xv*
Hessian fly, 45
Hidden Valley, Detroit, *xvi, xx*
Hieroglyphics, 57, 85-90, 95, 97-99
Hippopotamus, 13
Hoe, 87, 90, 93, 97
Hofstadter, Douglas R., 84
Holy Family, 92
Holy person, 114
Homeopathic remedy, 68
Horn Head, 3
Horoscope, 20
Horus, 40, 62-65, 89, 90-92, 96
Horus Temple, 62, 64
Housefly, 48
Hovering, 61
Hueco Tanks, *xii, xv-xx,* 99
Human body, 46
Human body, paramagnetic, 112
Human sacrifice, 50
Hummingbird, 42
Hydrogen, 116
Hypnotism, 46

I

Iberian Peninsula, 1
Idfu, 62, 63

Incarnation, 91
India, 68
Indian, American, 17
Indian cloak, 100
Infrared, 94
Infrared, coherent, 57
Infrared, film, 100
Infrared, radiation, 48, 57
Infrared, wavelengths, 59
Infrared, waves, 42
Insecticides, 117
Inundation, of Nile, 54, 93
Ions, atmospheric, 72
Ions, gaseous, 107
IR, coherent, 76
Ireland, 1-16, 26, 36, 98, 107, 117
Irish, monks, 108
Irish, sky, 8, 9
Iron, 23, 27, 58, 116
Iron filings, 30, 71
Iron oxide, 28
Isis, 89-92
Italian researcher, 35

J

Jade Mountain, 56
Jansky, Karl G., 21, 23
Jaundice, 68
Jefferson, Thomas, 17
Jesus, 55
Johnston, William, 25
Josephson Junction, 74
Joy in the brain, 69
Joy of Climbing, 69
Joy of Cooking, 69
Joy of Sex, 69
Jupiter, 19

K

Ka, 97
Kaolin, lime, *xviii*
Karmilov, V.I., 35
Karnak, 95
Karnak, boat, 61, 87, 88
Karnak Temple, 112, 113, 114
Keeneghan graveyard, 15
Keenaghan Lough, 9, 12, 15
Keenaghan Townland, 14
Kelp, 12
Kenneigh, 5
Kerry County, 12
Kestrel, desert, *xv, xvi*
Kevin, Saint, 11
Khartoum, Sudan, 93, 94
Kielbenna, 12
Kilconna, 12
Killinaboy, 7
Kilmacduagh, 7, 33

King's Chamber, 41, 53, 59, 61, 70, 112
Kite, black-shouldered, 42, 64
Klystron, 19, 23
Kom Ombo, 64, 95
Kom Ombo, temple, 88
Korean flag, 49
Krylov, A.V., 35

L

Lady Amhai, 90, 97
Lake Tana, 93
Lake Victoria, 94
Land, dead, 87
Langham, Derald G., 56
Language, secret, agriculture, 98
Laser, 17
Leaf, 82
Leigh-Mallory, George, 69
Levitate, 109, 116
Levitated loving persons, 117
Levitation, 50, 60, 61, 95, 111, 114, 115
Levitating, 61
Levitating force, 99
Levitating, person, 114
Libyan Desert, 54
Light, 76
Limestone, 22, 27, 58, 59, 70, 107
Limestone, bedrock, 98
Limestone, tura, 59
Limestone, white, 99
Lindbergh, Charles, 17
Little is a lot, 36
Loblolly pine, 117
Lough Beg, 12
Lough Erne, 3, 14, 34
Lough Neagh, 12
Love, 90, 97, 117
Lucky symbol, 56
Ludwig, Emil, 94
Luxor, 61, 62, 64, 87, 88, 95, 114
Luxor temple, 88
Lynx, 7
Lyra, 13

M

MacManus, Diarmuid, 38
Magic, 51
Magnaprobe, 59
Magnesium, 116
Magnesium sulfate, 30, 31, 32
Magnet, 27-29, 70, 71, 75, 103, 107
Magnet, bar, 75
Magnet, high-energy, 36
Magnet, rare-earth, 71
Magnet, support, 30
Magnetic antenna, 27, 29, 32

Magnetic charges, 72
Magnetic dipole, 75
Magnetic energy, 22, 71
Magnetic field, 23, 36, 72
Magnetic field, earth, 104
Magnetic monopole, 107
Magnetic place, 15
Magnetic resonance, 116
Magnetism, low-energy, 36
Magnetoelectric dipole, 105, 107, 109
Magnetoelectric dipole, gas, 104
Magnetosphere, 72
Magnite, 28
Manganese, 116
Manganese hydroxides, 58
Manhattan, 18
Map, 86
Maraini, Fosco, 111
Marconi, 17
Marijuana, 23
Maugham, William Somerset, 38
Maxwell equations, 71
Mayo, County, 12, 32
Medical circles, holistic, 48
Medicine, holistic, 46
Medicine, holistic view, 48
Meditating, 70
Meditative conditions, 108
Meelick, 9, 12, 13
Megaliths, 57
Memory machine, 36
Mer, 98
Mercenaries, 2
Mercury, 19
Mescalero Apaches, *xv, xvii, xviii*
Metal box, bilayer, 75
Metal, rubbed, 59
Metalworking, Celtic, 2
Meter wavelenghts, 221, 23
Mexican culture, 100
Mexico City, 50, 100, 101
Mica schist, 11
Microwave, 26
Microwave energy, cosmic, 25
Microwave radiation, 24
Middle dynasties, 88, 91, 95
Miracles, 100, 102
Miraculous image, 100
Mirror image, recursion, 78
Missionaries, Christian, 3
Model, Davis-Rawls, 103, 105
Model, fit, 110
Model, simple, 102
Model, tower, 23
Modeling, 22
Mode of motion, 43, 46
Modes, 32
Modes, energy, 36
Modulate, 64

Index   137

Moldboard, plow, 2, 15
Molecules of iron, 43
Molybdenum, 116
Monaghen County, 12
Monasteries, 4, 6
Monasteries, Buddha, 42
Monasteries, Christian, 26
Monastic expansion, 3
Monastic network, 3
Monks, Celtic, 12
Monks, Irish, 36
Monks, stone engineer, 25
Monopole, 74, 75, 81, 100, 103, 104, 105
Monopole, detector, 80
Monopole, magnetic, 74, 76
Monopole, north, 107
Monopole, north, magnetic, 116
Monopole, south, 107
Monopole, south, magnetic, 116
Monopole, tachyon signal, 81
Monopole, tachyon theory, 75
Monopoles, exist, 82
Monopoles, lightweight, 77
Monopoles, magnetic, 72, 77, 102
Monopoles, paramagnetic, 116
Monsoon, 93
Mosques, Moslem, 62
Mountain lion, *xvi, xvii*
Mouth, 97, 99
Mud brick, 55
Mud daubers, 57
Mud, fertile, 93
Mummified, 94
Mummy, 57, 87, 111, 114
Mummy, in pyramid, 39
Mummy, on water, 112
Mysteries, 51
Mystic, 107, 114
Mystic place, 14
Mystics, 87, 105, 115

N

Nahuatl, 50
Nasser Lake, 89
Nature god, *xix, 91*
Navaho culture, xviii
Navagation points, 8
Nelson, John, 18, 19
New York, 17
Newtown, Keenaghan Townland, 14
Nichols, Dr. Joe, 35
Night sky, 21, 23
Nile, 66, 87, 95
Nile, civilization, 87
Nile, Egyptian, 93
Nile River, 13, 54, 61, 93, 96
Nile, The, 94

Nile, Upper, 62
Nile Valley, 53, 92, 93, 98
Nileometer, 54
Nitrogen, 109, 116
Nobel Prize, 67, 102
Norman, 10
North monopole, 104, 108
North pole, of magnet, 103
North Star, 21, 66
Northern Ireland, 12, 13, 15
Norwich Cathedral, 47
NPK, 116, 117

O

Oak, 43
Obelisk, 56, 57, 108
Occult, 46, 92, 114
Oil lamps, 113
Okress, Ernst, 76
Old Kingdom, 55
Optical science, 114
Orbit, of earth, 13
Ordnance Survey map, 14
Organic chemical, 112
Organic content, of soil, 117
Organic matter, 91
Organic molecules, 107
Organs, scattered, 91
Orthorhombic cells, 78
Orthorhombic recursive network, 80
Orthorhomboidal recursive pattern, 81
Osiris, 85, 89, 90, 91, 92
Oughterard, 6
Owl's ear, 81
Oxygen, 107, 116
Ozma of Oz, 54

P

Pagoda, 40, 42
Pagoda, Chinese, 43
Pagoda, stone, in pyramid, 41
Pagoda-like structure, 57
Pah Kua, 56
Pantheism, 115
Paramagnetic, 23, 27, 28
Paramagnetic, antennae, 30
Paramagnetic, attractive, 36
Paramagnetic, body, 69
Paramagnetic, god, 92
Paramagnetic, HE force, 109
Paramagnetic, minerals, 46
Paramagnetic, rimrock, 98
Paramagnetic, rock, 69
Paramagnetic, round towers, 30, 32, 34
Paramagnetic, stone, 11
Paramagnetic, substance, 30
Paramagnetism, 59, 61, 94, 95, 107
Paramagnetism, in stone, 70, 100

Parish of Cairn, The, 14
Parthenon, 61
Particle-wave duality, 73
Peach tree borer moth, 44
Pesticides, 92
Petroleum, high energy, 37
Pharaoh, 50, 53, 57, 61, 64, 94, 112, 113
Phoenicians, 1
Phosphorus, 116
Photons, 100, 103
Photosynthesis, 82, 108, 109
Physical systems, 100
Physiological Chemistry and Physics, 74
Pictographs, *xvii, xviii*
Picture language, 50
PK power, 48
Planets, 19, 20
Plants, 104, 108
Plastic, dielectric, 107
Pless, Irwin, 46
Plexiglass, 22
Poem, *Wooden Begger,* 66
Polaris, 7, 13
Pole, north, 34, 35
Pole, south, 34
Pole star, 8
Poles, magnetic, 107
Poles, of magnet, 34
Political stability, 94
Polo, Marco, 43
Porphyry, 99, 107
Positrons, 107, 109, 110
Positrons, positive, 74
Potassium, 116
Prayer, 105
Precession, 13
Presman, A.S., 35
Priests, of Egypt, 115
Priests, paramagnetic, 111, 112
Priests, temple, 94
Princess, Egyptian, 14
Proceedings Royal Society of London, 75
Prophet, honest, 92
Psychokinesis, 48, 50
Puebloan culture, *xviii*
Punic War, Second, 1
Pylon, temple, 112
Pyramid, 39, 48, 58, 87, 94, 98, 99, 108, 117
Pyramid, Great, 8, 9, 24, 38, 40, 41, 50, 53, 56, 57, 62, 64, 70, 112
Pyramid, paramagnetic, 64
Pyramids, Egyptian, 13
Pyrenees, 1

Q

Quantum mechanics, 75, 103

Quantum particles, 77
Quantum physics, 72
Quantum theory, 77
Quartz, 58
Qutb Minar, 46, 69

R

Radiation, night sky, 25
Radio, 20
Radio, cosmic, 24
Radio emissions, 72
Radio waves, 21, 23
Radionics, 48
Radish planting, 49
Radishes, 35, 70
Rahad, 93
Ramisi, Professor, 61
Ram's Island, 12
Rare earth, 23
Rawls, Walter C., 34, 103
Rays, human health, 75
RCA, 18, 19
Re, 40
Reader's Digest, 40
Recorder, Rustrack, 78
Recursive network, 78
Recursiveness, 81
Red hematite, *xviii*
Red Rocks, Denver, *xix*
Red Tower, 42, 46, 69
Relics, of Buddha, 42
Religious orthodoxy of science, 102
Religious structures, 36, 42, 57, 59, 108
Resonance, 24, 71
Resonance circuit, 17
Resonant amplifier, paramagnetic, 73
Resonant cavity, 33
Resonant places, 15
Resonant stone, 50
Resonant waves, 15
Resurrected, 89, 92
Resurrection, 91
Rhomboidal shapes, 81
Right message, 25
Rio Grande, *xv*
River Erne, 14, 15
Rock, ground-up, 71
Rock, chart, 72
Rock, climber, 69
Rock crystal, 20
Rocks, and soil, 117
Rome, 1
Romans, 1, 3
Root plants, 34, 36
Roots, 35, 108
Roscam, 7
Rosetta stone, 97
Rotation, crop, 2

Index 139

Rotation, tillage-cattle, 2
Round towers, 3, 4, 5, 6, 9, 11, 14, 15, 16, 20, 21, 22, 26, 27, 30, 40, 48, 50, 57, 70, 98, 108, 109, 117
Round towers, carborundum, 30, 31, 35, 49, 70
Round towers, corrugated, 46, 48, 50, 57
Round towers, Irish, 59
Round towers, model, 60
Round towers, paramagnetic, 71
Round towers, resonance, 72
Round towers, sandpaper, 70
Round towers, smooth-sided, 46
Round towers, star plot, 7
Round Towers of Ireland, The, 4
Rubbing, 48, 59
Russia, 36
Russian literature, 36
Rye, 35

S

Sacred mountains, 108
Sahra esh Sharqiya, 54
Saint Brigid, 6
Saint Francis of Assisi, 117
Saint Ignatius of Loyola, 114
Saint Joseph of Cupertino, 115
Saint Peter's, 57
Saint Teresa of Avila, 109, 114
Salt, 23
Salt fertilizers, 92
Salunkhe, D., 35
Salvation, 91
Samarium cobalt, 71
Samarium cobalt, magnet, 28
Sand, 23, 95, 99
Sand, from east to west, 85, 87, 95, 96
Sand Creek, Denver, xx
Sandpaper, 22
Sandpaper, paramagnetic, 24
Sandpaper, round tower, 19, 23
Sandstone, 22, 27, 28, 99
Sandstone, paramagnetic, 24
Sandstone walls, corrugated, 46
Sarcophagus, king's chamber, 41
Saturn, 19
Sawdust, 23, 27
Scale, 99
Scarab beetle, 57, 58
Scent molecules, 57
Schist, 70
Schmidt, Barbara, 34
Science, 102
Scientific American, 76
Scientist, agricultural, 66, 67, 109
Scientist, materialist, 105
Scotland, 14
Scota, 14
Scotus, 14
Seed, 34, 117
Seed, sieved, 99
Seedlings, 35, 70
Semiconduction, 23
Semiconductor, paramagnetic, 58
Semiconductor, silicon, 22
Sensilla, 42, 46, 57
Sensilla, antenna, 58
Sensilla, arched, 45
Sensilla, corrugated, 46, 58
Sensilla, insect, 49, 59
Sensilla, pyramidal, 58
Sensilla, tapered, 43
Sensilla, wax-coated, 58
Sensitive people, 76
Septih, dog star, 99
Seth, 89, 91
Sex scent, insect, 76
Shabti, 85, 87, 96
Shannon River, 7
Shape, attraction, 71
Shape, of round tower, 49
Shapes, 109
Shapes, of religious structures, 42
Shelf life, 107
Shepheard's Hotel, 38
Shroud of Turin, 100, 102
Sick, 75, 117
Sick, people, 76
Silent music, 25
Silicate, 58
Silicon, 21, 27
Silicon, round tower, 24
Silicon carbide, 21, 23, 27
Silicon rectifier, 24
Silt, 54
Silt, paramagnetic, 89
Sin, 91
Singing spirits, xx
Skin temperature, 48
Sky map, 7, 16
Slave traders, 92
Slope of tower, 27
Smoked initials, xix
Soil, 36, 49, 50, 59, 95, 102, 108, 115
Soil, and rocks, 59
Soil, Egyptian, 93
Soil, erosion, 109
Soil, fertility, 3
Soil, living, 117
Soil, paramagnetic, 55, 64, 91, 92
Soil, rich, 91
Soil and Civilization, 94
Soil-loving people, 98
Soil solutions, 116
Solar cell, 21
Solar wind, 72
Solid-state detector, 75

140   Ancient Mysteries, Modern Visions

Solid-state physics, 20, 46
Solstice, winter, 8
Sotol cactus, *xv*
Source, paramagnetic, 99
South, American, 92
South, monopole, 104, 108
South pole, of magnet, 103
Southworth, George, 21, 24
Spain, 1
Speed of light, 77, 100
Sperrin Mountains, 12
Spin, north, 105
Spin, of magnet, 103
Spin, south, 105
Spine, corrugated, 42
Spine, insect, 36
Spine, plant, 36
Spine, sensilla, 76
Spirit of life, 64
Spirit of nature, 51
Square recursive plot, 82
Square wave, 77, 81
Square wave, monopole, 80
Standing stones, 45
Standing waves, 32
Stanford University, 77
Star map, 12, 15
Starvation, 4
Statistical range, 71
Static electricity, 48
Steel, 30
Steeple, 47
Steeples, cathedral, 42
Stem and leaf, 81
Stone, 107, 108
Stone, and soil, 104
Stone, force lines, 99
Stone, paramagnetic, 97
Stone and Bronze Age, 14
Stone building, 114
Stone of truth, 99
Stone rings, 57
Stone structures, 58
Stonehenge, 57
Stonehenge, England, 45
Stonework, 21
Strangler Dam, 89, 92
Striker, 79
Stroking, 48
Sudan, 92
Suleiman, Farag, 38, 39
Sulfides, 58
Sulfur, 116
Sun, 18, 20, 26, 72
Sun flairs, 104, 108
Sun god, 40, 72, 89, 90, 92
Sun rays, 75
Sunlight, 109
Sunshine charm, 68

Sunspots, 20, 103
Sunspots, activity, 107
Sunspots, flairs, 106
Sunspots, numbers graph, 104
Superconducting, 75
Superconduction, 17, 77
Superconductivity, 76
Superconductor, detector, 79
Superconductor, plant, 82
Supernatural, 51
Syenite-porphyry rock, *xiii*, *xvii*
Syllabic meaning, 97
Synagogue of the Holy Family, 92
Syria, 92
Szekely, E.B., 55

T

Tachyon, 74, 75, 78, 81, 82, 100, 102
Tachyon, magnetoelectric dipole, 75
Tachyon event, 83
Tarakanova, G.A., 35
Taste sensor, 47
Tepee, Indian, 27
Teflon, 107
Telekinesis, 59, 60, 61
Temple, Crocodile, 64
Temple, Egyptian, 62
Temple walls, 57
Temperature, 72, 104
Temperature, earth, 103
Teotihuacan, 50
Tesla, Nikola, 17, 18
Test, soil, 50
Thailand, 40, 42
Thebes, 88
Theory, or model, 117
Thondup, Prince, 111
Three Mile Island, 18
Thuban, 8, 9, 12, 13
Tibet, 111
Tibet, Secret, 111
Tielhard de Chardin, 105
Tillage-cattle rotation, 15
Tilling, 85, 87
Tilma, 100
Time-lag circuit, 81
Time lines, 78
Tomatoes, 35
Tombs, Egyptian, 40, 57
Tombs, frescos, 88
Tombs, west bank, 87
Tompkins, Peter, 61
Tower houses, 10, 98
Tranquilizing, 69
Trickle, time, 107, 109
Trinity, Egyptian, 91
Trinity, Holy, 40
Trinity, of forces, 92

Index 141

Tune into nature, 25
Tuning into nature, 22
Tuning into Nature, 77
Tuning pile, 33
Turlough, 31, 32
Tutankhamen, King, 88
Tyndall, John, 43

U

Ulster, 12
Unified crown, 89
United States, 36, 117
University, Irish, 11
University of Toronto, 78
Upper Egypt, crown, 89
Ursa Major, 7, 9
Ursa Minor, 55

V

Valley of the Kings, 64, 88
Vega, 8, 13
Velocity of light, 75
Venus, 19
Vespid wasp, 57, 58
Vibrating, 48
Vikings, 4, 24, 26, 48
Vikings, attacks, 3
Vikings, raids, 3
Virgin of Guadalupe, 100, 101, 102
Virgo, 7
Visible radiation, 57
Vision, mummy, 113
Visual prayer, *xix*
Vitamin C, 23
Vitamins, 36
Volcanic, 107
Volcanic, grains, 95
Volcanic, rock, 50, 93, 109
Volcanic, soil, 93, 98

W

Wall reliefs, Egyptian, 50, 64
Walls, temple, 113
Walters, Charles Jr., 2, 115, 116
Watchtower, 4
Water, and soil activity, 109

Water, balance, *xvi*
Water, diamagnetic, 91, 92
Wats, 42
Wattle huts, 6
Waves, paramagnetic, 59
Waveform, 71
Waveform, magnetic, 72
Waveguide, 23
Waveguide, dielectric, 57
Waveguide, radio, 71
Waveguide, tubular and rectangular, 22
Waveguide detector, 26
Wavelengths, centimeter, 28
Wax, 22, 107
Wax coating, 57
Wax slides, 42
Weed killers, 117
Weight, 99
West bank, of Nile, 87
Westminster Abbey, 39
Wheat, 35
White, J.E. Manship, 54, 91
White, Nile, 89, 92, 94
Wilson, Colin, 50
Wind, 72
Wind charm, 68
Windows, tower, 34
Winnie Callahan, 5
Witch's hat, 27, 28, 29
Witches' rings, corrugated, 45
Wood, Casey, *xv*
Wooden figure, 66
World War II, 3, 14, 66
Wright brothers, 102

X-Y-Z

X-rays, 20, 100
Yeats, William B., 12
Yin Yang pattern, 49
Young, Akerblad, 97
Young, Thomas, 97
Yoga, 114
Yugoslavia, 17
Zagazig, 53
Zinc, 116
Zodiac sky, 13

142 Ancient Mysteries, Modern Visions